A Guide to the Natural Environment of Lanark County

First printing, 1999
Revised edition, 2008
Revised edition, 2021

ISBN 978-0-9810462-0-4

Author's website: www.drpaulkeddy.com

Recommended citation:
Keddy, P.A. 2021. A Guide to the Natural Environment of Lanark County. (revised 2008 edition).

A Guide to the Natural Environment of Lanark County

Dr. Paul A. Keddy

www.drpaulkeddy.com

Revised edition
Janaury 2021

Revised edition
Janaury 2008

First edition
July 1999

Table of Contents

Preface

This guide to the natural environment of Lanark County is now moving into its fourth decade. It has also entered a new era—now it can now be purchased online and delivered directly to your home. Of course, where possible, you should still support local book stores, but not all have kept the book in stock, and I have heard repeatedly from potential readers who could not find copies. So now there is another option—copies can be delivered to you in only a few days.

This new printing has been lightly updated. We corrected a few typographical errors, which otherwise, for many years, have been dutifully hand-corrected by Cathy Keddy. Now we can put away the whiteout and black pen. Thank you, Cathy, for your precision and effort. I have also simplified the title, which many people found to be a mouthful. Yes, Lanark County's ecosystems are produced mostly by geology, flooding, and fire (hence, "Earth, Water, Fire" in the former title) but it does not need to say so in the title. Also, the familiar line drawing of gray tree frogs has migrated away from the front cover (thank you Howie Coneybeare) to be replaced by an image that better conveys the subject matter of the guide. Gray tree frogs are wonderful wetland animals, and their calls are delightful in May and June, but there are many other equally remarkable wild species here.

We have learned more about the county over the past decade. For example, field work for the planned Ontario atlas of reptiles and amphibians has allowed us to update the range of Five-lined Skinks and Gray Ratsnakes in Lanark County in Map 18 (page 53). Terence Currie's book on the great fire of 1870 has allowed us to update Map 10 (page 28) Another important landmark has been continued inventory of all the wild plants of Lanark County by David White (www.lanarkflora.com). There are new conservation groups active in the county, including the Mississippi Madawaska Land Trust, so the section "Some Other Sources of Information and Guided Tours" has been updated with new information (page 69).

The important natural areas in our county are still not adequately protected. New information has allowed us to better understand the location of our most important natural areas. This provides an essential foundation for environmental planning in Lanark County. Research on species at risk that has been carried out at the national and provincial scale has updated our understanding of the status of rare and threatened species in our own county. For example, all of our species of turtles are now considered species at risk. The Cerulean Warbler is in decline across North America and is now considered to be endangered. The Monarch Butterfly is also in decline across North America and is consider to be endangered as well. The Gray Ratsnake is now considered to be threatened on the Frontenac Axis. Extinction of wild species is something that is happening here and now and is not just a phenomenon that happens in distant places like Africa and South America. The absolute first priority to protect our species at risk is to protect the wild places where they live.

For this reason, this latest printing of the guide includes a new appendix listing 30 of the most important natural sites needing protection in Lanark County (Appendix A). I also added five guidelines for future conservation planning in the county (Appendix B). All of the 30 listed places should be regarded as core natural areas in the county. All urgently need protection, including buffer zones to surround them. These 30 areas also need to be linked to one another with natural corridors. This is a high priority challenge for citizens who are concerned with the natural environment of Lanark County. Many of these areas, particularly the wetlands, store large amounts of carbon and so are essential in efforts to reduce the impacts of global warming.

Interest in outdoor recreation continues to grow. This short guide to Lanark County is not intended to introduce you to individual walking trails, but rather to explain the natural environment of the county as a whole, to help users appreciate what they are seeing along the county's many trails. New guides to specific trails are emerging, such as *A Guide to Hiking Trails in Ottawa and Region* by Victoria Walsh (www.girlgonegood.com). It is my hope that this new printing of my Lanark county biography will guide you to fuller appreciation of our outdoors overall and encourage you to act to ensure that the wild beauty of Lanark County survives for future generations.

Dr. Paul Keddy, January 2021, www.drpaulkeddy.com

Acknowledgements

Although part of my youth was spent in Lanark County, I remained ignorant of many of its important features. I would never have learned to see the landscape in the way this book portrays without the early encouragement of my parents, Norm and Jean Keddy, and later tutoring by Bruce McBride, Dan Strickland, Tony Reznicek and Dan Brunton. Many people have helped me locate the information presented here. I thank Dan Brunton, Paul Catling, Don Cuddy, Jeff McNaughton, Cathy Keddy, Martin Streit, David White, and the University of Ottawa Map Library Staff for their assistance. The Friends of Algonquin Park graciously gave permission for reproduction of the line drawings of animals by Mr. Howard Coneybeare, and Rochelle Lawson assisted in preparing the other drawings. I particularly thank Chris Drummond for his work in preparing the maps, and Andrea Million for her work as my assistant. Special mention must also go to Marie Elmsley for her help in finding the funding to hire Chris Drummond and Andrea Million; the book would still be only partially completed without her enthusiastic support. Ian Keddy prepared the maps and designed the revised edition. While many more aspects of Lanark County's ecological profile remain to be described, these details will have to wait for a future edition. Feel free to write with your suggestions for the next edition or to advise me of any errors that may have crept into the manuscript.

Table of Maps

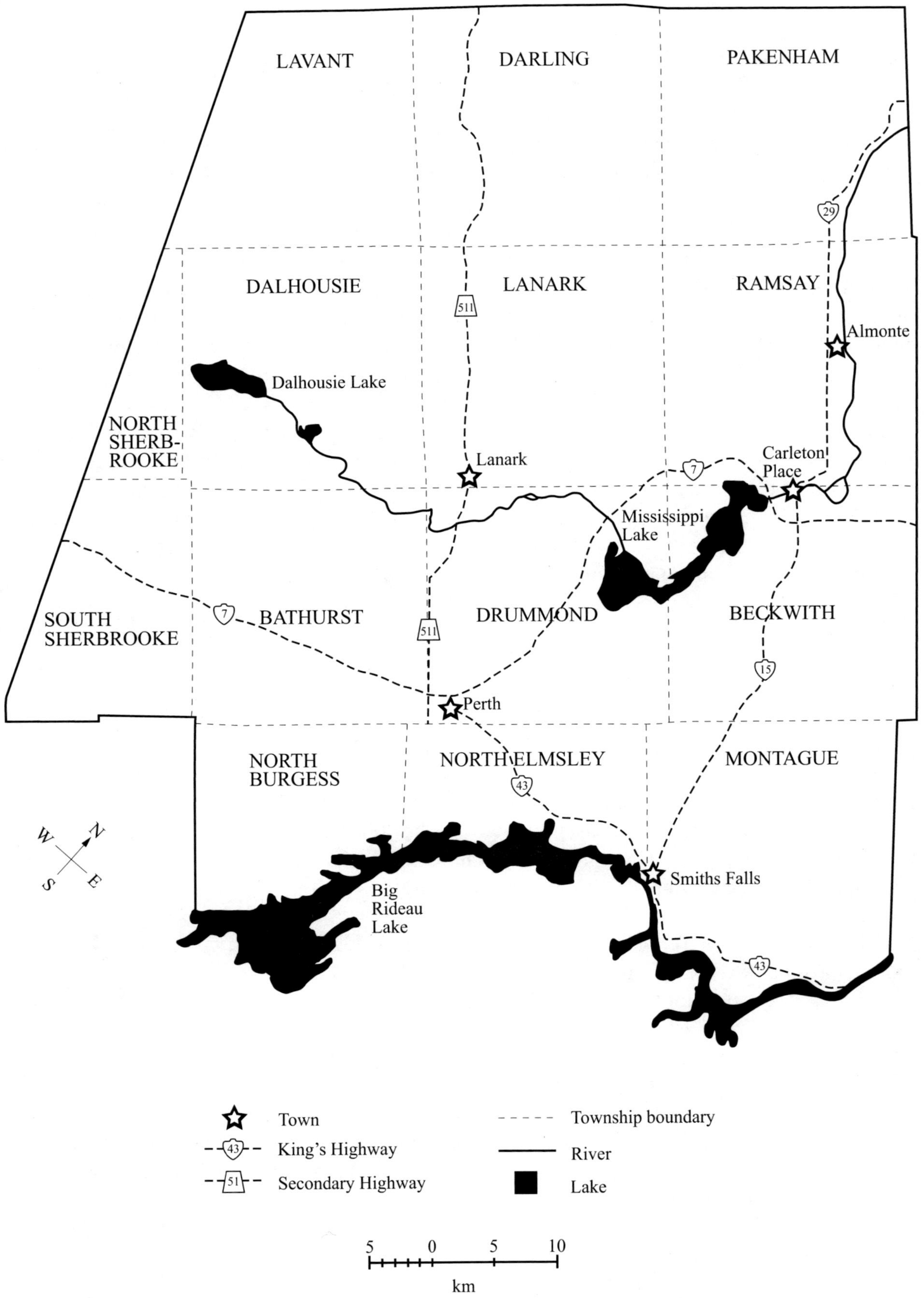

Map 1. Some modern human features of Lanark County (see Map 19 for subsequent boundary revisions).

Introduction

Map 1 shows Lanark County as we normally see it — highways, cities, and surveyed townships. This tells us how one species — human beings — recently placed its mark on the county. Almost all of these lines reflect events of the past 200 years of settlement. We humans are recent arrivals here, and the regular grids laid out by early surveyors hide the natural patterns that are more real than any survey line.

In 1998 many of the familiar towns and townships shown in Map 1 were amalgamated into a smaller set of municipalities (see Map 19, p. 74). Going clockwise from the top right these are (1) Mississippi Mills, (2) Montague, (3) Drummond and North Elmsley, (4) Tay Valley, and (5) Lanark Highlands. In addition, there are three smaller areas designated as separate towns: Carleton Place, Smiths Falls, and Perth. Since it will take some time for us all to adjust to these changes, and since all the data used in this book, including historical references, were collected before this amalgamation, the names of the old townships are used throughout.

Geology

A tale of two bedrocks[1]

The bedrock gives Lanark County a split personality. Probably the single most important factor in the natural and human history of Lanark County is the predominance of Precambrian rocks of the Canadian Shield (Map 2). The Canadian Shield is made up of some of the most ancient rocks on Earth — some close to 4 billion years old. It occupies more than 4.5 million square kilometres of Canada. Granite and gneiss are the most common rocks here; both are very hard and produce thin and acidic soils. The shield prevented the spread of agriculture and was an obstacle to the construction of roads and railroads. How the early settlers must have cursed their luck as they heaved huge boulders across their fields and onto accumulating stone walls. How they must have envied their neighbours only a few miles away who were lucky enough to have a limestone plain, or a pocket of deeper soil covering the shield. The rolling landscape we associate with much of Lanark County takes its form from these hard rocks. Although millions of years of erosion, including at least four recent ice ages, have smoothed the land a little, the rugged rock outcrops are a constant reminder of shield country. It will be here long after our human civilizations are gone.

The southeastern portions of the county are another story. These rocks are much softer and younger — a mere quarter of a billion years old (give or take a few million years one way or the other). The fossils in them tell us they were formed in an era when shallow oceans covered eastern North America. They are composed of sandstones and limestones. For simplicity, we will call this area a limestone plain, but there are several different kinds of limestone here as well as sandstone. Because limestone plains are flat, it makes it much easier to construct roads. Just notice how straight Highway 7 is east of Carleton Place, and how much it starts to wind as it climbs onto the shield just west of the town. Limestones are relatively soft rocks, and therefore also tend to form deeper soils. Owing to the chemistry of the rock, the soils which form are also

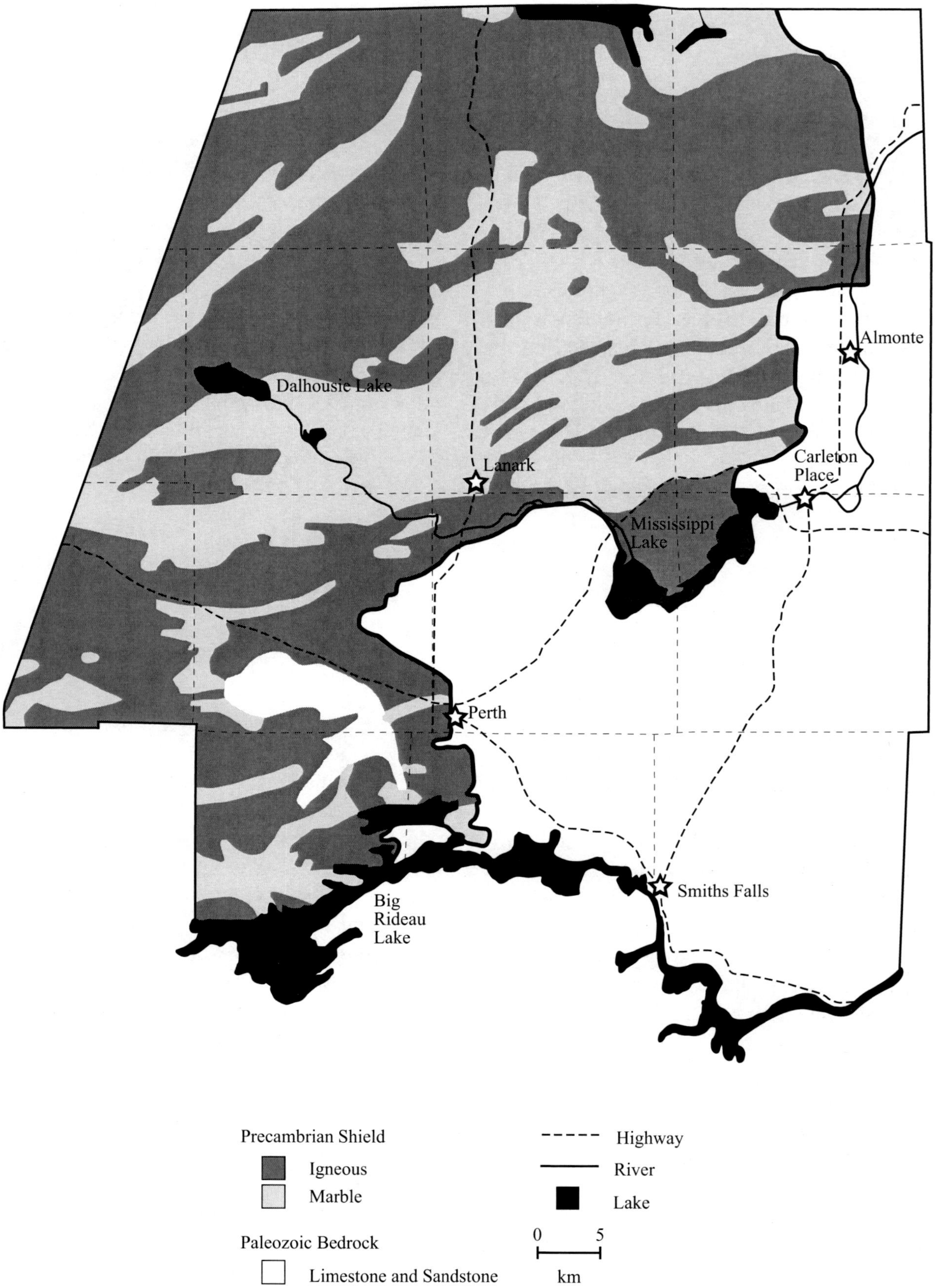

Map 2. Distribution of the Precambrian shield and Paleozoic bedrock in Lanark County.

neutral or alkaline. Anyone who has purchased lime to improve their soils would appreciate the situation of farms built right on limestone. Certain kinds of plants are so sensitive to the acidity of the soil that they are found only in areas with limestone soils. You can more or less map this for yourself by watching the rock cuts along the edges of roads — layers of greyish-white rock usually indicate limestone. The distribution of buildings built of limestone is also a good guide — the main street of Perth testifies to the fact that it is on the edge of a limestone plain.

The importance of being marble

So the county is divided into two regions: Canadian Shield, and limestone plain. Rocky, acidic soils predominate in the former; deeper more alkaline soils in the latter. But marble complicates this, and makes Lanark County unusual. The areas of marble are shown in Map 2. Marble is relatively uncommon over the rest of the southern Canadian Shield, but here in Lanark County we have some large areas of it. Marble is formed from ancient limestones which have been altered by heat and pressure. This produces a strange kind of hybrid landscape. It looks like the typical Canadian Shield — rocky, rolling hills. But the soils are alkaline and more fertile. Plants which one expects to find only on limestone start to appear. Having marble produces different kinds of forest, better soils for agriculture, and unusual wetlands. Marble will come up as a topic again because it is so basic to understanding the county.

But before leaving the topic of bedrock, one more thing must be said about marble and acid rain. Thin acid soils typical of the Canadian Shield are highly sensitive to acid rain. Deep soils over limestone are not. Thus, rocky areas northwest of us like Algonquin Park are highly sensitive, whereas clay areas southeast of us are not. In Lanark County itself, we would expect the shield areas of Map 2 to be more sensitive to acid rain than the limestone plain. But marble throws this pattern off. The marble areas are much less sensitive than the surrounding granite and gneiss rocks. And within the limestone plain, the sandstones around Perth, Smiths Falls and west of Mississippi Lake are more sensitive to acid rain than most limestones. So if you want to know whether your particular property is at risk from acid rain, you need quite a detailed map. In general, if you are on marble or on the limestone plain, sensitivity to acid rain is moderate to low. Elsewhere acid rain is a significant threat to soils and vegetation.

The ice age and the Champlain Sea[2,3]

The most important event in the history of Lanark County was the melting of the great ice sheets that once covered the area. The last ice sheet melted a mere 12,000 years ago. The ice age was a comparatively recent event — there were humans around to see it, and most of the plants and animals familiar to us today were alive then. Of course, since Lanark County was covered in ice kilometres deep, our familiar animals and plants lived elsewhere much farther south.

The ice age also caused most of Lanark County to disappear under water. There were two reasons for this. First, the weight of ice was so great that the land sank beneath it. Second, the melting glaciers produced immense volumes of water.

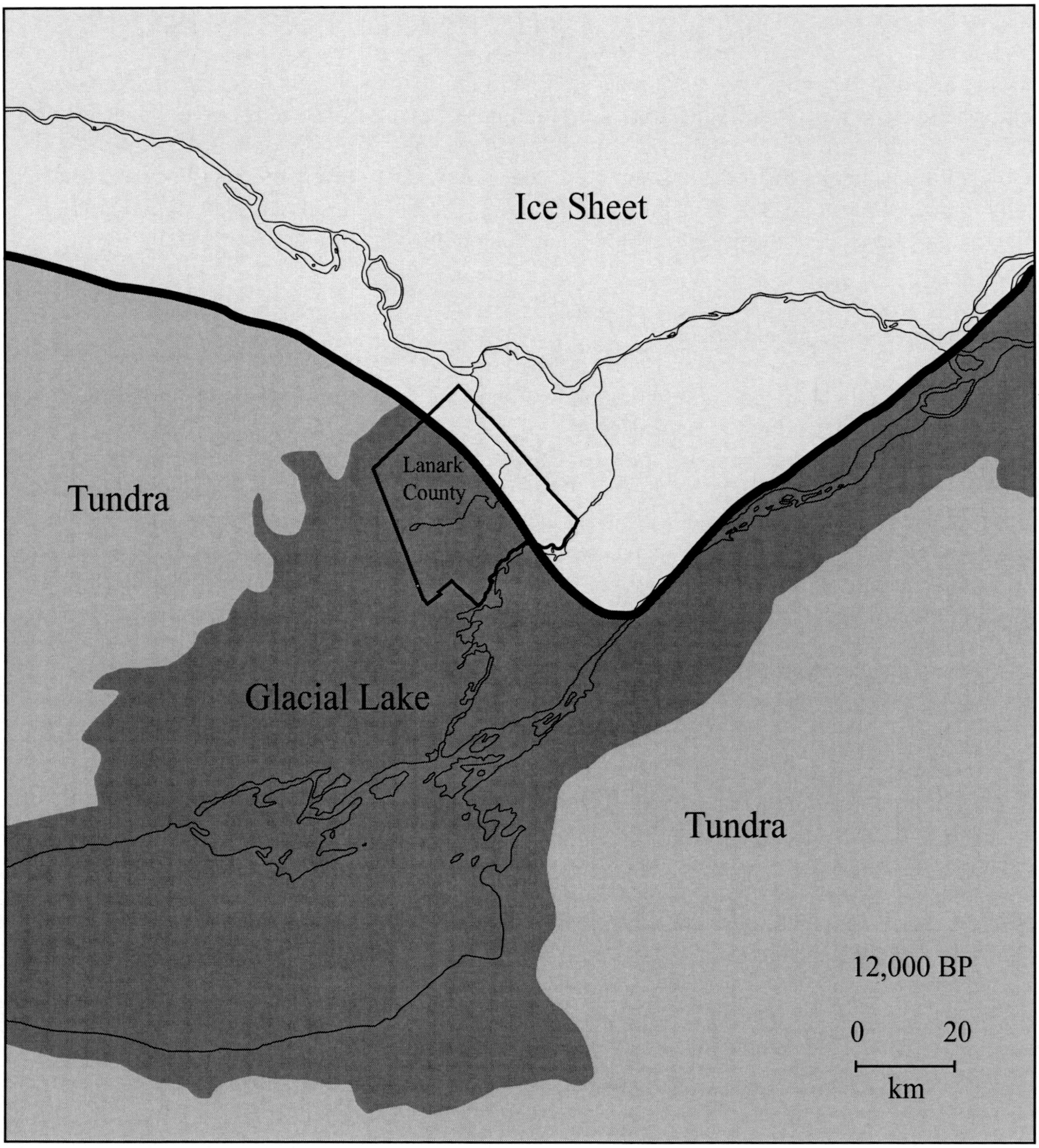

Map 3. Extent of the glacial lake in eastern Ontario at 12,000 BP.

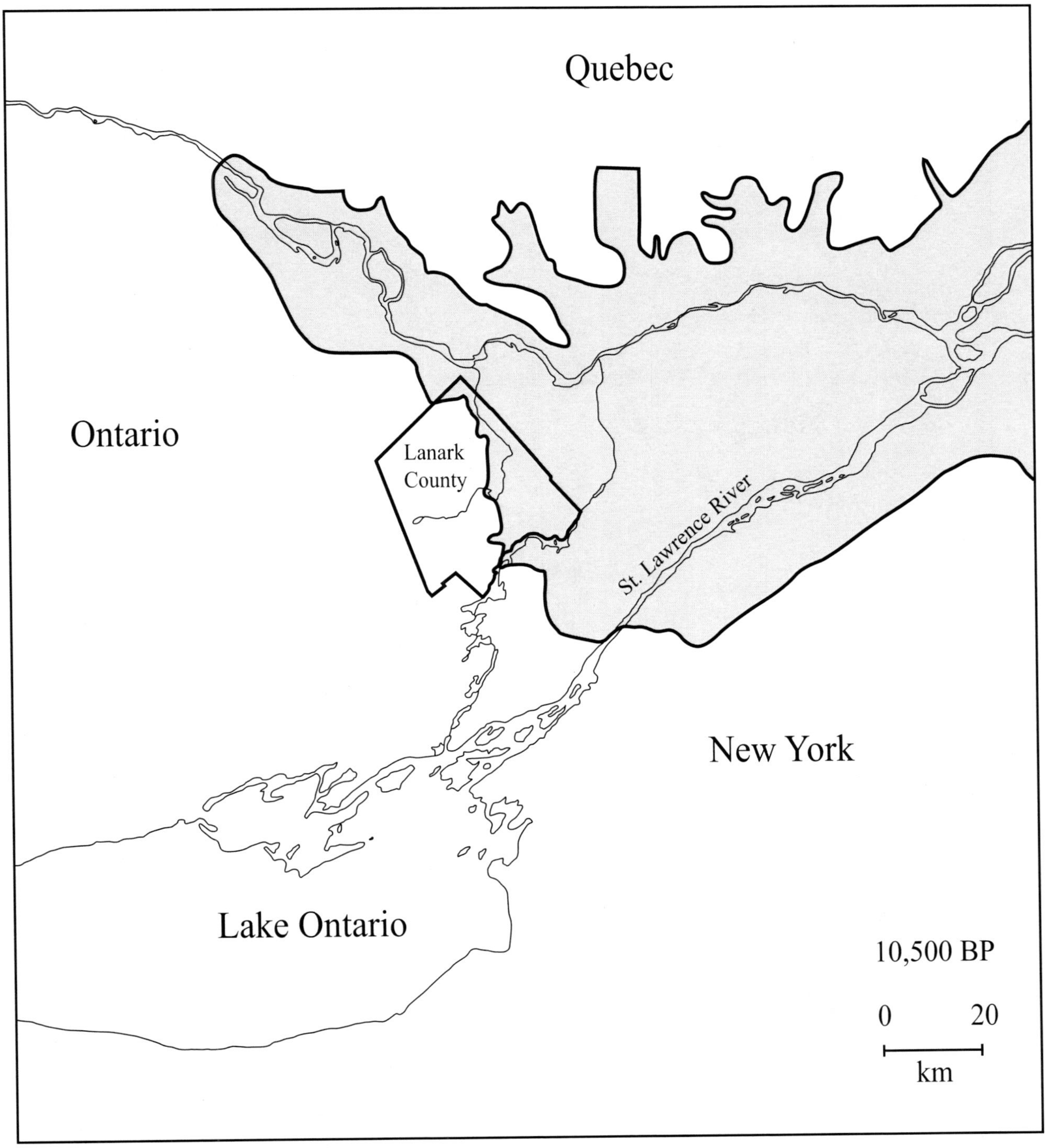

Map 4. Extent of the Champlain Sea in eastern Ontario at 10,500 BP.

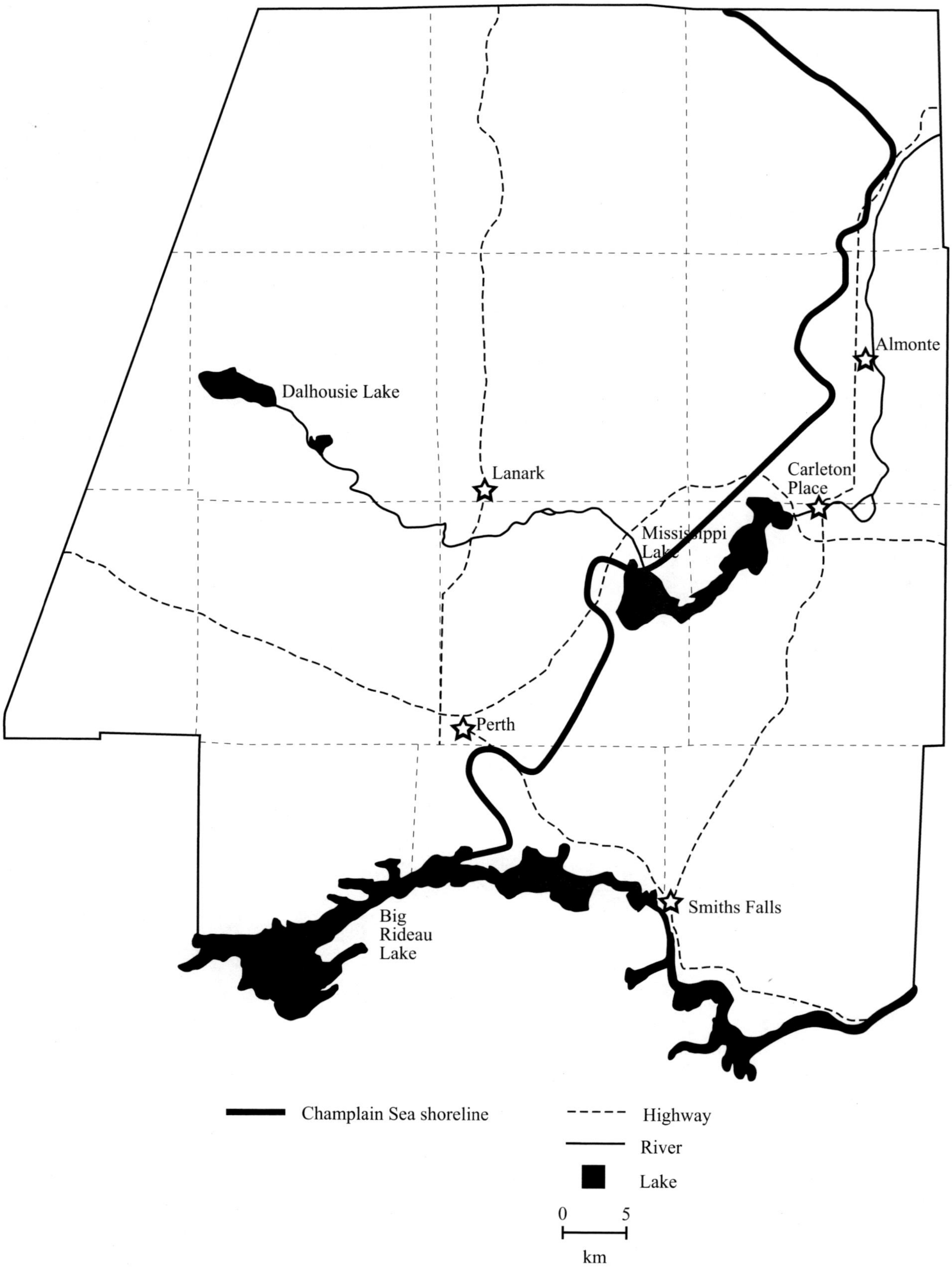

Map 5. Shoreline of the Champlain Sea in Lanark County, adapted from Currie, T.M. 2009. The Ottawa Valley's Great Fire of 1870: The Nineteenth Century Press and the Reality of a Great Disaster. Creative Bound International, Canada.

About 12,000 years ago southwestern Lanark County had melted free of ice, but a huge wall of ice still stretched northward from the St. Lawrence River through modern day Smiths Falls and Carleton Place (Map 3). Drainage eastward was blocked by this ice sheet. As a consequence, a huge lake formed. Some of the sticky clay soils which fill many of our valleys likely date from this period.

Once the ice melted farther north, the lake could drain to the sea. But the land was still so low from the weight of the previous ice cover that the sea flooded in from the Gulf of St. Lawrence and formed the Champlain Sea (Map 4). The shoreline of this sea stretched roughly from Perth to White Lake; features of this old shoreline can still be traced today (Map 5). Every time one drives along Highway 7 from Carleton Place to Perth, sections of this old shoreline are crossed. One might have been able to see whales swimming and seals basking on the shores of Drummond and Beckwith townships. The sea only lasted about 2,000 years, but its presence is still felt today. More clay plains were laid down under this ocean. In many places, one can find clay plains that have been cleared for agriculture interspersed with Precambrian knobs that have been left in forest. These clay plain and rock knob landscapes create the odd shaped fields that are familiar in some parts of Lanark County.

Geologists who study such events have found clay deposits where there is a rapid transition from freshwater clay to marine clay; these document the change from the freshwater lake to the salt water ocean. Whether the large clay plain west of Perth was part of the ocean, a freshwater bay, or an isolated lake, is still not clear. Ongoing studies of the Perth Bog by scientists at the Geological Survey of Canada may clarify this.

Hills and valleys from the ice age

The present day landscape of Lanark County (Map 6) is largely the result of such ice age events. Many areas of the shield and the limestone plains have shallow soil because as the glaciers slid south, everything was pushed before them and deposited in the United States. The farmers there are undoubtedly grateful for having received our soil, and there seems to be little we can do to get it back. But how the early settlers who cursed their rocky land would have longed for what the glaciers took away.

In a few areas the glaciers left deeper deposits which now cover the bedrock. West of Perth along Highway 7 is a large flat area of clay plain. These clay plains were once covered by shallow water from the melting ice; fine clay particles settled out. Although these soils are sticky and often poorly drained, they are better than rock for farming. These areas were both cleared and drained by settlers, making it difficult to determine what the original forests were like on them. Smaller pockets of clay occur throughout the township. Many are still fields but the drainage ditches through them remind us that these areas were once swamps.

Some depressions were swamps for so long that peat began to accumulate. Peat is formed when waterlogging prevents decay. Over the short term (thousands of years) dead plants accumulate as peat. Over the long term (millions of years), these peat deposits might turn to coal. Large pockets of organic soil are only in the southeastern parts of the county, but many smaller depressions can be found scattered through the shield country and on the borders of lakes.

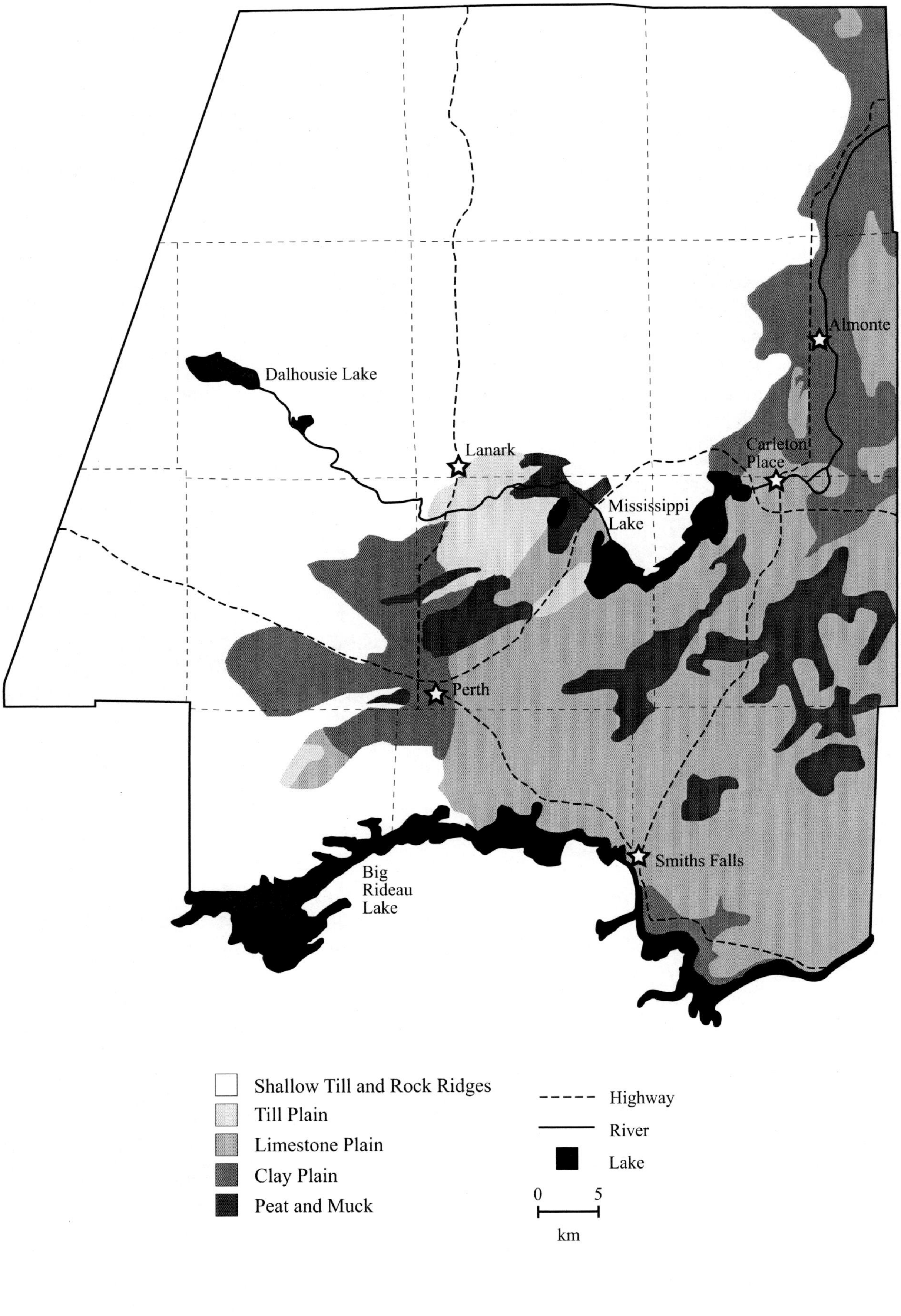

Map 6. Physiography of Lanark County.

So here are the four main features: two kinds of exposed bedrock (shield or limestone) and two kinds of wet depressions (clay or organic soils). This describes most of Lanark County. A notable exception is the till plain in the very centre of the county where the receding ice sheet generously left a deeper deposit of gravel and sand, and deeper soil developed. This pocket seems remarkably out of place when one first enters it by driving north of Highway 7 between Perth and Carleton Place. One suddenly enters gently rolling land with deep soil and prosperous farms — it looks just like a small piece of southern Ontario transported into the middle of Lanark County. Lucky were the settlers who got their farms here. Another smaller piece of this landscape can be seen along Highway 7 where it passes by the south end of Mississippi Lake.

Map 7. The distribution of Temperate Deciduous Forests of North America.

A County of Trees

Overview

Lanark County was once covered in dense forest. One could have entered on the east side, have walked for days to cross it, and have never seen the sky except through a canopy of leaves unless one waded across a river. A squirrel, they say, could have travelled from Montreal to Toronto without ever touching the ground. Those days are long gone. The forests that remain have been devastated by logging, wildfire and clearing for agriculture. The story of our forests really is little different from that of the rest of eastern North America, with two exceptions: our soils are thin and rocky in many areas, and parts of the county were never settled for agriculture.

Let us familiarize ourselves briefly with the forests of eastern North America before we consider the details of the county. A few short centuries ago eastern North America was covered by deciduous forests.[4,5] Similar forests covered western Europe and eastern Asia. These forests were remnants of the great Arcto-Tertiary forests which for millions of years covered much of the Northern Hemisphere. These original forests are largely gone. Remnants persist in three forms: extensive areas of second growth recovering from past logging and agricultural practices, tiny fragments thought to represent fragments of virgin forest, and the occasional larger protected area. In Lanark County we have no large protected areas, and no known remnants of virgin forest. The early loggers were too efficient. There are, however, some areas of maturing forest, and some of these may approximate the early forests in certain limited ways. Some of them may return to their original form if protected.

The best remaining deciduous forest in North America, and quite possibly the best in the world, occurs in Great Smoky Mountains National Park in the United States. It is impossible to fully appreciate Lanark forests without visiting the forests of the southern Appalachians. It is stunning to see mile after mile of hill sides covered in mature forests, with our familiar sugar maple and hemlock trees growing mixed with unfamiliar southern species such as tulip trees and species of magnolia.

Lanark County lies near the northern edge of deciduous forest in North America (Map 7). Owing to the severity of the climate, there are only a few common tree species — a fraction of the number that occur in the southern Appalachians. The commonest tree species in a mature stand of forest would be sugar maple, American beech and basswood. Others found in lesser abundance would include red oak, bur oak (often mistakenly called white oak), white ash, bitternut hickory, butternut, yellow birch and hemlock. The kinds of trees depend to some extent on the soil. Drier and rockier sites might have more red and bur oak, along with evergreen trees such as white pine. Cool north facing slopes will tend to have more hemlock and yellow birch.

Farther to the north the Boreal Forest stretches from east to west across Canada. True boreal forests are much farther north than Lanark County, but we are far enough north that some sites support stands of white spruce and white pine. However, most of these evergreen forests are second growth — that is to say, they are filling in old farmland or burned areas, and probably do not represent the original forests of the area. Given several more centuries, these forests would probably revert back to deciduous trees.

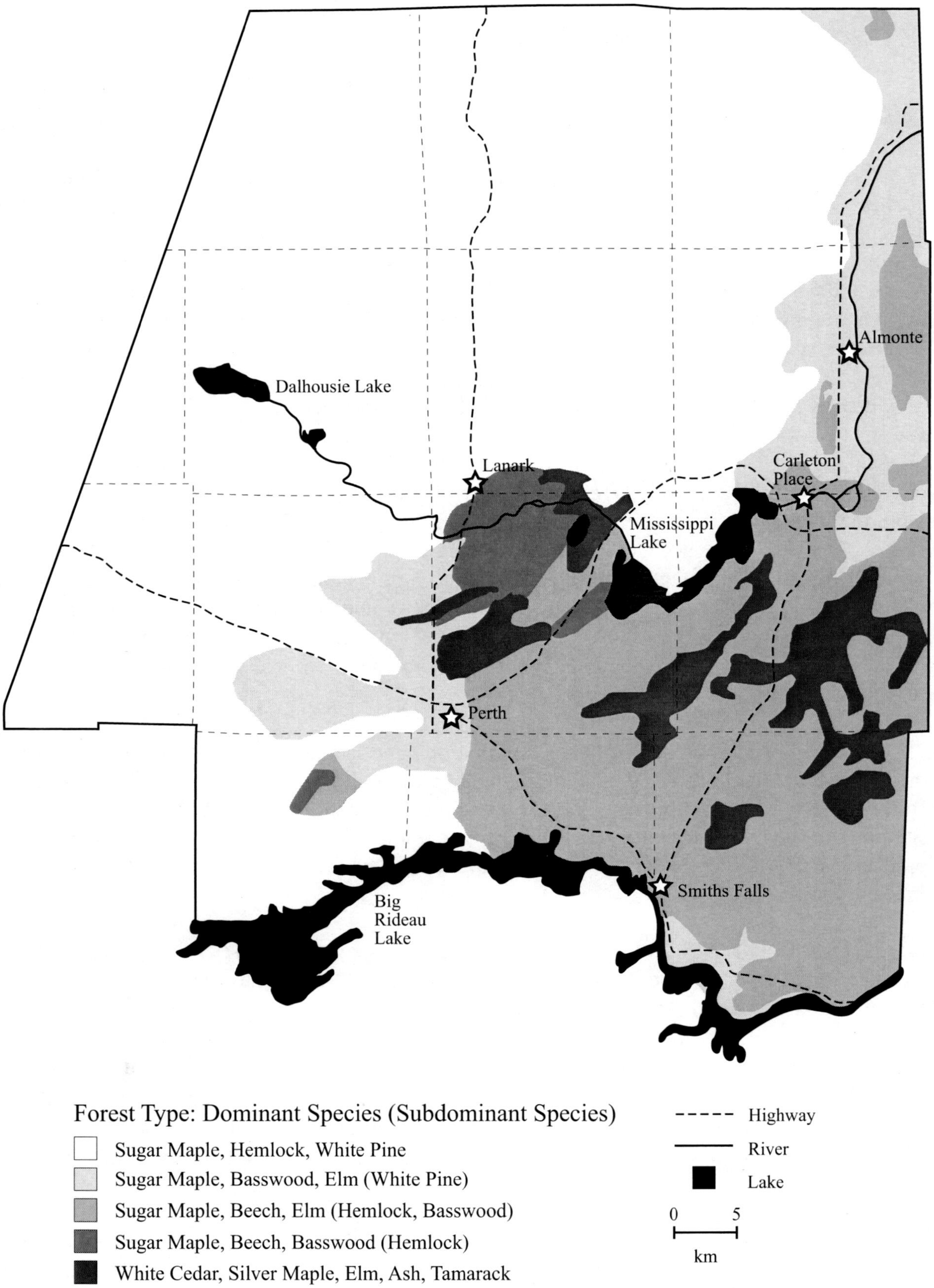

Map 8. The original forests of Lanark County.

The forest primeval

Map 8 shows the main forest types that occupied Lanark County prior to the arrival of settlers. At that time, sugar maple, white pine and hemlock were probably the three most common living things in Lanark County. There is a good deal of educated guess-work in drawing such maps. This one is based upon early surveyors' notes, as well as the types of soils shown in Map 6.

Early settlers in Beckwith Township described their experience of these forests as follows:

"*There is something in the ponderous stillness of these forests — something in their wild, torn, mossy darkness, their utter solitude and mournful silence which impresses the traveller with a new aspect each time he sees them ... In Upper Canada the endless hills of pine give way at last, or at most stand thinly intermingled with gigantic beeches, tall hemlocks and ash, with maples, birch and wild sycamore, the underwood of these great leafy hills. Mile after mile, and hour after hour of such a route was passed — a dark black solitude, with here and there a vista opening up, showing the massive trunks, grey as cathedral ruins, which bore the rich canopy of leaves aloft.*"[6]

Sycamore does not grow here; early settlers often did not know the correct names for trees. Another immigrant said:

"*The greater part of the forest, the underwood or bramble, is not as thick as at home .... all grown over with large trees, some fresh and green, others half rotten and a great many rotten from top to bottom, and almost as many lying in all directions as are standing, with not a living creature to be seen or heard except a bird or two, and the owl screaming in your ears at night.*"[7]

Another Lanark County pioneer described a forest "*through which the sun never penetrated.*"[8]

There is little scientific data available on the original forests of Lanark County.[9] Most of what we know is inferred from fragmentary historical records, the study of existing remnants, and reference to preserved pollen in lake sediments. It is therefore often difficult to measure the changes that have occurred over the past few centuries. None the less, there are obvious trends and changes in selected species. The changes in these species tell us something about the forests as a whole. Only a few generations ago Passenger Pigeons were observed to cover tree branches entirely and darken the sky in large flocks. Early settlers told of felling thirty or more at one shot. Now they are extinct. Some large mammals have been extirpated from Lanark County although they survive elsewhere — these include the wolverine, timber wolf and eastern cougar. Birds that exploit disturbance, such as the Chestnut-sided Warbler, or the ubiquitous House Sparrow, have increased in relative abundance, while birds requiring mature tracts of forest such as Black-throated Blue Warblers have probably declined. The average size of trees has certainly declined, and species that require disturbance (poplar and birch) or tolerate it (ironwood) have increased in abundance. The abundance of almost all spring flowers has plummeted, owing both to forest clearance and grazing within forests. Ginseng, once common, was reduced by clearing and grazing of forests as well as by over-harvesting; it is now on the province's rare plant list. Wild leeks are now declining for similar reasons. Such species not only require mature forest, but they reproduce very slowly since they produce only a few seeds each year, and often take nearly a decade to grow large enough to reproduce. They therefore take human generations to recover from disturbance. These sorts of changes in plants and animals indicate that our forests have changed too.

The shallow rocky soils in the shield country would have been dominated by hemlock, sugar maple and white pine. It appears that hemlock and white pine have declined over the past centuries. On deeper soils, sugar maple was again a dominant, but beech and American elm were more commonly mixed in. The elm trees were almost eliminated just a few decades ago by a disease introduced from Europe; one can see bleached dead trunks, and the occasional living tree, but it is now hard to imagine that they could have been a dominant species. beech has also apparently declined. It is slow to reproduce from seed, the seeds are heavy and do not disperse far, and it may be that birds like Passenger Pigeons once played an important role in spreading their seeds. In some parts of Lanark County, it appears that beech trees indicate areas of forest that have been relatively less disturbed. The clay plain forests are least well known, in part because the clay plains have been so thoroughly cleared for agriculture. We therefore must depend largely upon the surveyors' notes to describe them. It appears that sugar maple was dominant, but it tended to be mixed with basswood, elm and white pine. Finally, in wet areas, swamp forests had developed. These were probably dominated by white cedar, silver maple, ash, elm and tamarack. All of these species, with the exception of elm, are still very common in wet areas.

Some information concerning the appearance of forests in Montague Township was provided by a settler faced with the task of forest clearing, "*There is always a great quantity of timber lying on the ground, which has been blown down by the wind, and must be cut up.*"[10]

The fact that two of these early quotes refer to fallen dead trees is noteworthy. There is growing evidence that standing and fallen dead wood is important for many species of wildlife. For example woodpeckers need dead trees as a source of food, and many other birds nest in rotten trees. Salamanders may breed in ponds, but they retreat to fallen dead trees for much of the year. Some species of wildflowers such as mitrewort grow largely on fallen trees. Even other trees such as hemlock and yellow birch tend to grow as seedlings on rotting logs. Today we tend to think of fallen dead trees as waste; in fact, removing dead trees may be one of the most damaging things we can do to forest wildlife.

Map 9 shows present day forests. Several obvious differences stand out. First, the area of forest has been significantly reduced, particularly in the southeastern parts of the county. Second, the forest has been broken into fragments. In the southeast, we could describe the landscape as cleared land with scattered islands of forest, whereas in the northwest we could call it forest with scattered clearings. The kinds of trees in the forest have also changed. The remaining forests, particularly in the southeastern areas, are often composed of species that thrive with, or at least tolerate, human disturbance. Areas of white birch, poplar, white spruce and white cedar generally testify to relatively recent disturbance.

Our best forested areas — the areas of mature forest that come closest to the original forests of Lanark County — are in the northwestern areas. Regrettably, no comprehensive maps are available showing our areas of oldest growth, even though active logging is still occurring in these areas.

The factors controlling forests

To understand the present day patterns in our forests, it is necessary to start with an understanding of the original forest patterns, and then examine how humans have changed them. Altogether we have four main factors: (1) geological history, (2) water availability, (3) humans and (4) fire. Let us consider them in turn.

(1) Geological history[11,12]

During the last ice age there were obviously no forests here at all. Trees survived in refugia far to the south. As the ice sheet receded, species began to migrate north. By studying pollen grains in lake sediments, biologists have been able to map the northward movement of different tree species. In general, the story goes something like this. The glaciers and the Champlain Sea left bare rock, sand, gravel, and clay. The early plant communities occurred in open tundra-like conditions with willows, grasses and sedges intermingled with spruce trees. Birch, alder and juniper were also common. In general, herb tundra was replaced by shrub tundra which was replaced by a spruce and poplar woodland. As the climate warmed and trees moved back from the south, pines and hemlock increased in abundance. In time, deciduous forest trees such as sugar maple, beech, basswood and oak arrived and became the dominant trees in the landscape. The types of bedrock and soil then determined the distribution of these tree species within the county.

(2) Water availability[13]

All plants need water to grow. The geology of a site determines how much water is available to plants. Rocky hill sides and sand plains allow water to run off and therefore such sites experience several months of drought each summer. Clay soils and depressions, in contrast, can accumulate water, and too much water cuts off oxygen to tree roots. Some trees are sensitive to drought, and therefore cannot live in the dry sites; other trees are sensitive to flooding and cannot grow in wet sites. Figure 1 therefore shows three main forest types (although professional biologists recognise many more special variants of these).

Dry sites

Dry sites such as south facing rocky slopes, tops of hills and sand plains have larger proportions of white pine and red oak. Because they are dry, these sites are particularly likely to burn from time to time, and so many of the species in such forests also show various adaptations to tolerate fire. The understorey species on such sites often include blueberries, large-leafed aster and bracken fern. In very dry locations, and with more fire, large openings can form; these clearings can become prairie-like and support unusual plants and animals. Prairie Warblers and Five-lined Skinks (Lanark's only species of lizard) may then occur.

Mesic sites

Sites with deeper soil and more moisture develop sugar maple-beech forests. We may think of this as the typical forest type in Lanark, with the other forest types being variants of it produced by extremes of flooding, fire or human disturbance. These sites support a rich array of forest plant species such as white trilliums, bloodroot, wild leeks and many orchids and ferns.

Many of the flowering plants have to flower and grow in the few short weeks of spring before they are shaded by the trees.

Wet and flooded sites

Clay plains and depressions have trees that tolerate various degrees of flooding. Red ash, black ash, red maple, yellow birch and elm were likely the dominant trees here. It is difficult to know what the original wet forests were like because these sites have often been intensely disturbed. Slightly higher areas that are dry for part of the year can have a rich ground flora, but in lower areas plants such as foamflower occur only on fallen logs which provide a partial refuge from flooding. Trees such as yellow birch or white cedar also grow on the fallen logs. Many wet sites have been drained for agriculture. Others have been flooded by beavers and turned into marsh. The spread of Dutch elm disease in the 1960s was another shock in remaining areas of wetland forest. Elm was once a common species, and as the elms died, they created large gaps which admitted sunlight and allowed new species to grow. Most wet forests are still recovering from the death of the elms.

River floodplains and some clay areas in valleys are flooded or waterlogged for so long that even species tolerant of wet sites cannot survive. In such habitats one finds silver maple swamps. These, as the name suggests, are dominated by one tree species, although white cedar, ash and elm are often mixed in. The rich wet soils support primarily sedges and ferns — ostrich fern (source of commercial fiddleheads) and sensitive fern are the most common species.

(3) Humans

First Nations

The first wave of immigrants arrived in North America between 20 and 30 thousand years ago. They crossed over the Bering Strait and gradually spread south and east over North and South America. We know little about their biological effects on forests — with a few notable exceptions.

At one time North America had a remarkable group of large mammals that rivalled those of Africa.[14,15] This megafauna, as it is called, included mammoths and giant ground sloths. Some palaeontologists have observed that their decline to extinction appears to coincide with the arrival of these first humans, and have therefore suggested that early hunters were responsible for the extinction of this entire fauna. Other biologists then pointed out that such large mammals may have had important impacts on forests, ranging from grazing to seed dispersal. Would we have had more kinds of trees in Lanark County if there had been giant ground sloths to carry their seeds back north after the ice age? Is this an effect of early human hunters? While this is only speculation, we may postulate that human impact on our forests can be traced back to far before Europeans arrived.

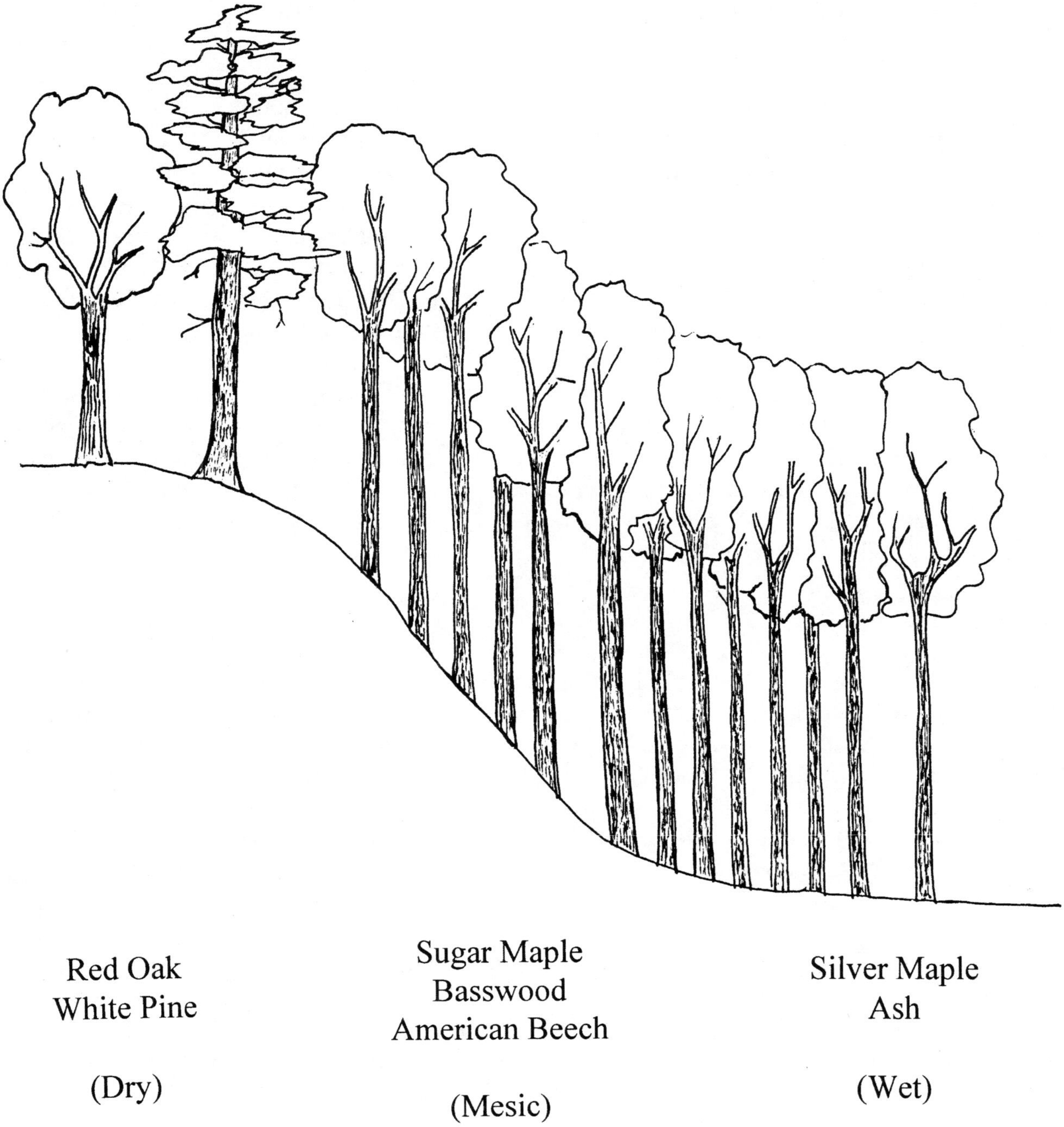

Figure 1. The forests of Lanark County typically change from pine and oak on rocky ridges to ash and silver maple in valleys and swamps.

Little is known of the aboriginal populations in Lanark County, since both archaeological and historical data are limited.[16,17,18] In historic times the Algonquins were the main group who lived here. Farther to the south, the Iroquois and Hurons had agricultural economies based on corn, beans and squash. Significant areas of land were cleared around their settlements. We can say with some confidence that there was little clearance of land for agriculture by the Algonquin Indians, although there is some evidence that in the 17th century Algonquins did create small fields which were planted with corn, beans and squash. In general it appears that the economy was based upon hunting and gathering, with trapping and trading becoming more important as European demand for furs resulted in the fur trade. The Matouweskarini, an Algonquin band, were settled along the Madawaska River in the early 17th century; farther to the east the Onontchataronon were settled along the South Nation River. Iroquois raiding parties apparently travelled up the Rideau to attack tribes living along the Ottawa River.

In summary, prior to the arrival of Europeans, Aboriginal populations here were probably small, with hunter-gathering economies. Their environmental impacts were probably fairly small — the creation of small clearings, possibly an increase in fire frequency, and the trapping of fur-bearing animals. Their ancestors may or may not have already eliminated mammals such as mammoths and giant ground sloths.

Europeans

All this changed with the arrival of Europeans with their hunger for resources such as furs, minerals, and wood. We can only touch on the highlights here. The first Europeans explored Ontario in the 1600s, and were followed by fur traders. Small numbers of primarily French and British settlers came during the 1600s and 1700s. The pace of development increased with the arrival of about 10,000 United Empire Loyalists fleeing the American War of Independence in the 1770s. In 1800 Philemon Wright started a small farming community on the site of present day Hull, and by 1826 Bytown had sprung up on the site of present day Ottawa. The Rideau Canal was built between 1826 and 1832 to link Bytown with Kingston.

Logging companies were already at work in the Ottawa Valley, and the pace of cutting quickened during the early 1800s when the Napoleonic Wars blocked access to traditional British timber sources in the Baltic area. white pines were the first to go. They were felled, squared, floated downstream, assembled into large rafts and eventually floated to Montreal for export to Europe. Waves of loggers then removed the smaller timber for sawmilling. In a little over a century the great forests were largely gone. Although we may romanticise the history of life in the lumber camps, historian Grant Head, who has studied forest exploitation in 19th century Ontario, was led by his studies to observe "*... the clearing that resulted from 19th century sawlogging must have shocked even 19th century minds, and can today be described only as a desecration.*"

Along with loggers, waves of settlers poured into eastern Ontario.[19,20,21,22,23] Lanark County was colonised from the southeast to the northwest. In some cases loggers were loath to see forests cut and simply burned when they could produce timber to feed sawmills. But on the other hand, small farms produced food to support the logging camps. The best land, that on limestone, was settled first. Beckwith Township, for example, was surveyed and settled in the

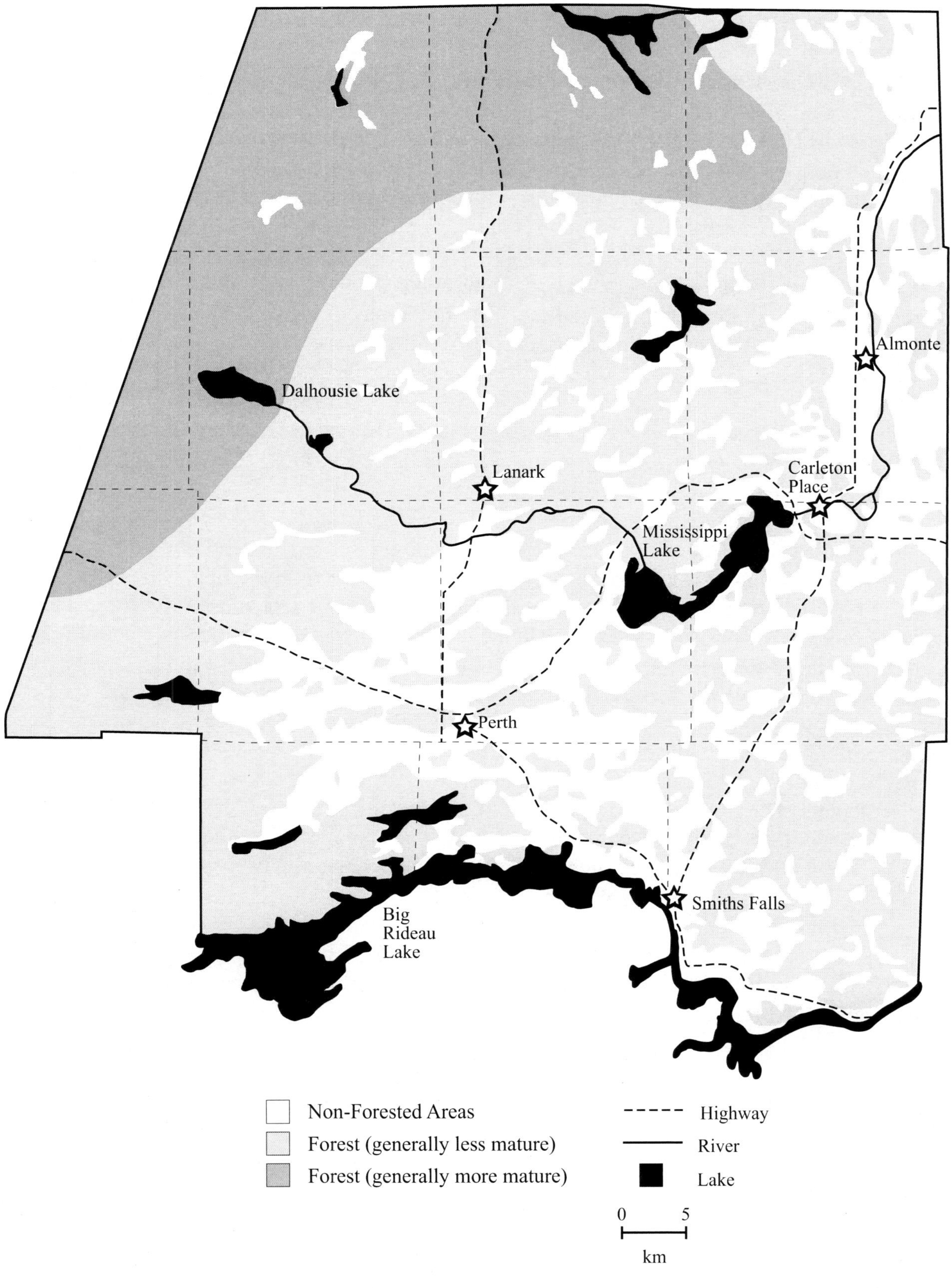

Map 9. The present distribution of forests in Lanark County.

1810s. A map of the District of Bathurst published in 1824 already shows all of the townships of present day Lanark County, as well as the settlements of Perth, Lanark and Richmond[24] (Map 4.16 in *Ontario's History in Maps*). Once the best land had been settled, settlement roads were extended in a network along the edge of the Canadian Shield; in 1856 the Hastings Road was opened northward into the shield country of south-central Ontario, and the Mississippi Road extended east into the northwestern corner of Lanark County.

What was the motivation behind trying to farm the shield? It was obvious to some that these rocky soils could not support agriculture. The limitations of the shield had already been described by surveyors beginning in the 1820s, leading one observer to conclude "*The general quality of the land is extremely rocky and broken,* so *much* so *indeed, that, in a district of about five hundred square miles, not a portion, sufficient for a small township, could be contained in any one locality, of a generally cultivable nature.*"[25] (J. W. Bridgeland cited by Gentilcore and Head 1983, p. 79). With hindsight, we may just see this scheme as another boondoggle, not unlike so many modern ones.

Railroads were also built to take advantage of and promote settlement, logging and mining. The shield country was penetrated by the Kingston and Pembroke line, which was chartered in 1871.[26] Owing to the difficulty of the terrain, it never extended past Renfrew. It took 12 years to build 112 miles. Contrast this with the CPR which was built in the same era; it was extended from the Great Lakes to the Pacific Ocean in only ten years (1875-1885)! A principal function of the K& P railway was to carry pine and other lumber southward to Kingston for export to the United States. Lavant, Folger, Clyde Forks and Flower Station were some of the settlements along the K& P. In so far as the environmental history of Lanark County goes, the principal impact of the line was the increased destruction of forests through logging, fire and human settlement. The K & P Railway was taken into the CPR system in 1913. It was eventually abandoned as a railway and the tracks removed. Now it is a popular recreation trail. The K& P lasted only about 40 years, and the trains and tracks may now be only history, but it may take 400 years for the surrounding forests to recover.

It is hard to imagine today what the landscape may have been like at the turn of the century. We have little data from Lanark itself, but a fine example from a neighbouring geographical area with similar soils and bedrock. By 1913 a special commission was set up to report on the state of forests in the Trent watershed in Peterborough County.[27] C.D. Howe reported in 1915 that logging and repeated burning had led to massive soil erosion, observing "*one frequently finds stumps of trees from one to two feet in diameter on bare rock ... The trees could not have germinated and lived for many years on bare rock.*" (One can find stumps like this today in Lanark County, testifying to the soil now eroded into the valleys and streams). Wildfire was frequent "*... there are no places in the former pineries, outside of the swamps, that have not been burned at least once since lumbering.*" "*Thirty thousand acres of the region examined have been burned over many times.*" One of his main conclusions was that "*... the principal effort should be to devise a proper forest policy for the area.*"

A summary of some important events and eras in the history of Lanark County

Confederation only occurred in 1867. By 1889 maps show a network of railways and canals criss-crossing eastern Ontario and linking Lanark County with other centres such as Parry Sound, Toronto and Montreal. Roughly a century later the Ottawa-Hull metropolitan area has become the fourth largest in Canada with a population size of about a million people — the same number of people that occupy all of Saskatchewan or Manitoba. The transition from forest, to farm, to suburbs has occurred in just a few human generations, or about the life span of a sugar maple tree!

Fire in the forest

Fire has a natural role to play in many kinds of forests. Some tree species such as yellow birch and hemlock actually may require light ground fires to create the conditions for their seedlings to establish. Other species such as white pine and white birch establish after more severe burns. We do not fully understand the role of natural fires in Lanark County. As a first approximation, it is likely that drier areas of shallow rock and thin soil burned in the order of once a century. But humans have greatly modified these natural patterns. First there were extensive wildfires in the 1800s. Some of these were the result of slash left by logging; others were deliberately set to clear land for agriculture. Historical records refer to smoke obscuring the sky for days. It may have been not unlike the photographs we now see of burning rain forests in Brazil. As mentioned above, C.D. Howe did a thorough study on the effects of fire in Peterborough County.[28] This is of great interest to us because the rock and soil types are not that different from Lanark County. He showed that each time a site was burned, the proportion of hardwoods (sugar maple, basswood, beech) decreased, and the proportion of aspen and pine increased. After more than three burns, only poplar and white birch remained.

As a result of such reports, the Department of Lands and Forests (now the Ontario Ministry of Natural Resources) invested a great deal of effort in fire suppression. We have probably all seen signs and brochures about Smokey the Bear, but putting out all fires is also a mistake. We went from one extreme to the other. First there was a surge of fire in the 1800s that was undoubtedly much higher than normal; then a reduction in fires in the 1900s to a level that was probably much below normal. The first would have led to a surge in the establishment of pine, birch, poplar and oak; the latter probably reduced the establishment of the same species along with yellow birch and hemlock.

One of the most dramatic fires was the Great Fire of 1870, which apparently began in northern Lanark County and then swept eastward toward Ottawa from the 17th to the 23rd of August (Map 10). Historical accounts provide an overview:

"*The summer of 1870 was practically a long drought. No rain had fallen for two months and the land lay parched under a sky of brass. Creeks and water courses shrivelled to trickles or were baked to seamed beds of clay .... There seems to be no definite origin to the fire, or rather series of fires, that spiralled from such widely separated sources in the townships of Fitzroy, Huntley, Gloucester, Nepean and in Hull across the river.*"[29]

"[The fire] *burst like an escaped monster out of the woods; it seemed to break spontaneously out of the ground and swept with fearful velocity over the parched fields. The punk dry fences acted*

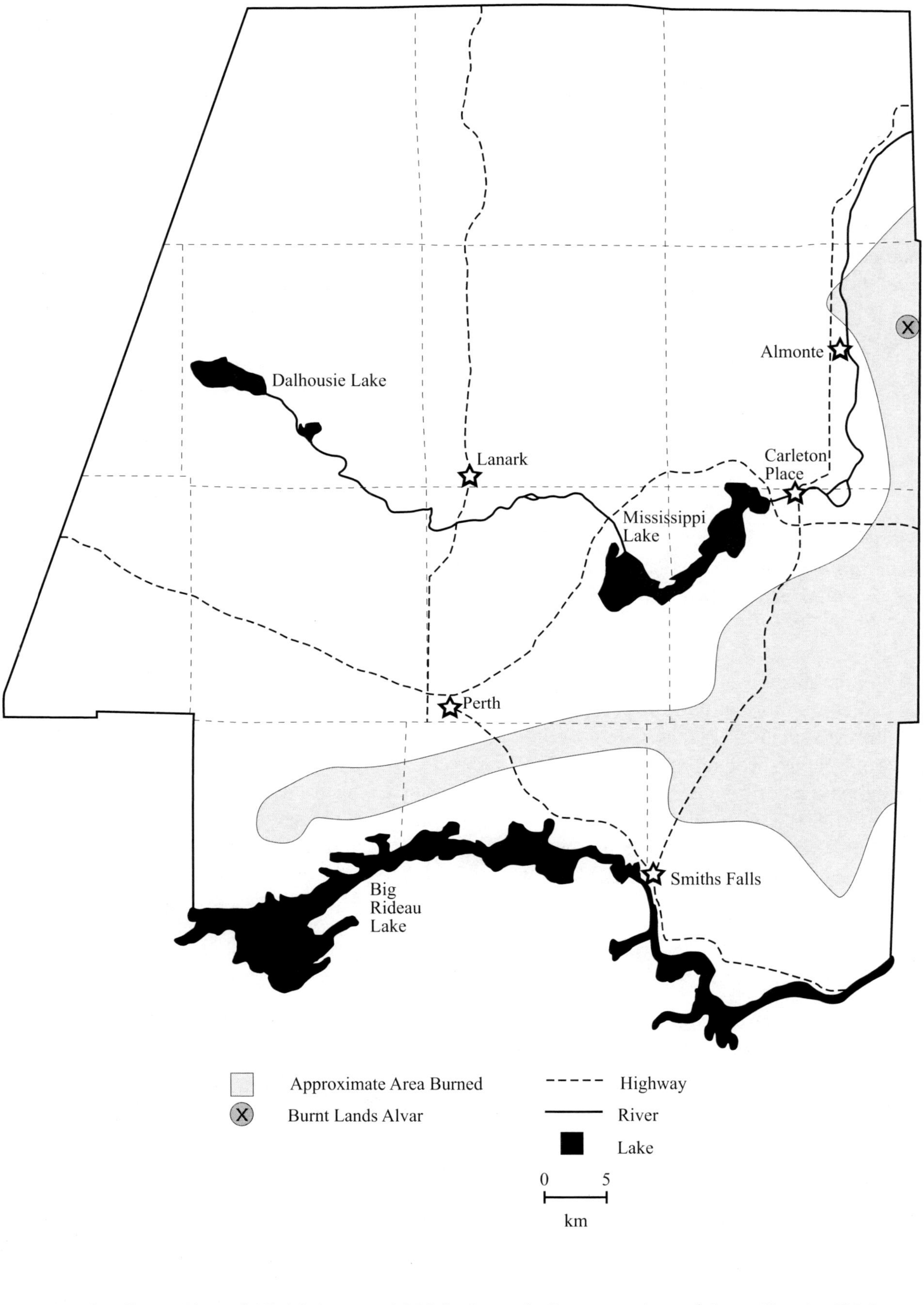

Map 10. The Great Fire of 17-23 August 1870 in Lanark County, adapted from Currie, T.M. 2009. The Ottawa Valley's Great Fire of 1870: The Nineteenth Century Press and the Reality of a Great Disaster. Creative Bound International, Canada.

like powder trains, and invariably conducted the fire to the farm buildings. When barns took fire, the scene was one of fearful grandeur, the whole sky looked like a vast curtain of flame; from the burning barnfuls of hay and grain, great sheets of fire were carried through the air intact."[30]

The view from Parliament Hill was threatening. " ... *the smoke from the ascending flames darkened the horizon and during the day the view on every side was bounded by heavy banks of cloud.*"[31]

"*On Richmond Road alone there were over two thousand people homeless and without sustenance, save what they could pick up; numbers were making their way to the city, their consternation and grief being terrible to witness.*"[32]

When it appeared that the city of Ottawa would itself be engulfed by the advancing flames:

"*Firemen, soldiers and citizens then went into action to save their own city from imminent destruction. But the deciding factor which stopped the wall of fire was the Dow's lake reservoir of water, unleashed by the cutting of the Lewis dam which flooded a wide area to the Richmond Road.*"[33]

Map 10 shows the approximate area of the fire. We need to remember that natural fires have been an important part of forest ecology in North America for thousands of years. Many of the forest species in Lanark County remain adapted to a natural fire regime. The Burnt Lands Alvar (p. 57) was probably not produced by this event. Many of the rare plants found in this alvar indicate its origins lie some ten thousand years earlier — in the tundra-like vegetation that occurred after the ice age. The Great Fire may, however, have changed the extent of the alvar. Overall, it is possible that settlers increased the extent and intensity of the fire with piles of brush and dead trees produced by efforts to clear the natural forests for agriculture.

The significance of our forests

Because there are so many trees around us, it is tempting to take them for granted. They are important for many reasons. Some reasons are purely economic. We heat our homes with them. They keep our mills running and fellow citizens employed. They attract tourists who support hotels, stores and restaurants. Other reasons are equally important but rather more difficult to measure with dollars. Forests maintain the water levels in streams and rivers. They produce oxygen for us to breathe. They remove carbon dioxide from the atmosphere to help control global warming. They produce wildlife for hunting and trapping. They provide wildlife for humans to simply appreciate. And they provide a refuge for growing numbers of rare and endangered species.

But as history shows, our forests have been under assault for over two centuries, with large areas gone and most others degraded. It is shocking that our forests have been so badly managed that local mills are closing. We obviously have to take steps to manage our forests better. This means (1) preventing further loss of forest, (2) planting new forest to replace that which has been lost, (3) restoring existing forest so it is better able to support our economy and (4) setting aside some areas for complete protection so that our children can see what the great forests of Ontario once looked like. The first three steps are so obvious that little more needs to be said about them. The knowledge now exists to manage our forests properly; the question is whether there is sufficient will to do so. One wonders how many more mills will have to close before the message is understood. The box on page 32 summarises some basic principles for the wise management of forests.

More can be said about part (4) — setting aside some protected areas. Around the world there is a growing movement to set aside protected examples of each world ecosystem. From the rain forests of Brazil to the tundra of Canada, an international system of protected areas and parks is growing. There is more or less world-wide consensus that at least 12 percent of the world's surface area should be set aside in parks. The remaining 88 percent of the landscape should be more than enough to grow our food and feed our mills. Although great progress is being made in some parts of the world, including Ontario as a whole, eastern Ontario and Lanark County fall well below this figure of 12 percent. The protected forest acreage in Lanark County is limited. Only two areas of upland forest are protected: those in Murphys Point Provincial Park, and those in the Darling Township Forest ANSI. There are several areas of swamp forest that are protected as significant wetlands. These are discussed in more detail under the topic of Special Places.

We therefore might ask whether it is still possible to participate in this global effort. Understand right at the start that there appears to be no true wilderness left in Lanark County. Although parts of Lanark County are frequently referred to as 'wilderness', this is misleading. Wilderness refers to areas that have been little-disturbed by humans. Nearly all forested areas show evidence of human disturbance. Having said this, the northern parts of the county are the least disturbed (Map 9). Moreover, there are areas of forest here that are still very much better than what is left almost anywhere else in eastern Ontario. Therefore, even though there may be no true wilderness, Lanark County has some of the best forests left in eastern Ontario; they may not be old growth, but they are oldest growth.

There is one other important consideration. Lanark County has one type of forest that is uncommon elsewhere in the province — deciduous forest growing upon marble. It is only in the past few years that biologists have begun to appreciate just how important this is, and that Lanark County is unique in this respect. So this is something our community can contribute to conservation that few other areas can match. If we selected carefully, forests could be permanently set aside to represent forest over marble, gneiss and granite. There are large areas of Crown Land that are already owned by the public and perhaps a future edition of this book will be able to include a map of protected forests.

Porcupines feed on maple, beech, oak and hemlock trees. They can been seen at any time of year, and most often at dusk.

Some Principles of Restoration Forestry

The objective of restoration forestry is to recreate the original forests typical of Lanark County. Not only does this style of forestry enhance the environment for native plant and animal species, but it creates forests with the high quality saw logs and veneer trees essential for the future of local industry.

1. The cut is always light, and the forest canopy is never opened more than about ten percent. Cutting is planned to create small gaps similar to those created naturally by fallen trees.

2. Forests are gradually converted toward their original composition. In upland sites, desirable trees typically include sugar maple, beech, hemlock, white pine and red oak. Special attention is paid to rarer hardwoods such as bur oak, white oak, butternut, elm and hickory.

3. Dead fallen wood (called 'coarse woody debris' by forest ecologists) is left to rot whenever possible to provide habitat for fungi and salamanders.

4. Standing dead trees are left as wildlife trees for woodpeckers, nesting birds and small mammals such as flying squirrels.

5. The best trees are left to provide seed.

6. Trails are kept narrow and shaded, and allowed to regrow with native plants.

7. Large trees are always present to maintain shade and provide habitat for canopy dwelling wildlife.

8. Large clearings are avoided, since these encourage weedy species rather than native forest species.

9. Deer populations are kept low to minimise damage to regenerating trees, whereas fishers, black bears and other native mammals are protected.

10. Each piece of forest is connected to other forest permitting wildlife to move easily. Forested strips are planted or allowed to regrow to connect isolated islands of forest. Roads are kept narrow.

11. Cutting and skidding are restricted to winter months when the ground is frozen and there is snow cover. Horses are used when possible.

12. All human activity is kept to a minimum in May and June, as this is the period when the soil is soft, spring flowers are growing, frogs are breeding and birds are nesting.

Great Blue Herons forage for fish and frogs along the edges of lakes and ponds. They nest high up in trees, often in large colonies approaching a hundred birds.

Birds

The mixture of forest, wetland and fields results in a remarkable diversity of bird species in Lanark County. There are breeding records of species ranging from hummingbirds to vultures. Some parts of the county have more than 100 kinds of breeding birds — more than you can find in many areas of Ontario farther south. Although we often assume that southern regions of Ontario have more kinds of birds, the forest and wetland areas have been so reduced that some bird species no longer occur there. Bird enthusiasts have mapped the breeding distribution of each bird species in Ontario based on two five-year periods of field work (Map 11).[34,35]

The distribution of each kind of bird depends upon the availability of food, shelter and nesting habitat. Different species have very different requirements. Some eat seeds, some eat insects. Some nest in trees, some nest on the ground. Some nest only in deep forest, others only in open fields. It is hard to generalise about our more than 100 species, except to say that a majority of our birds depend upon forests and wetlands.

Changes in the forests have changed the kinds of birds we have today. We have already lost one forest bird forever. Flocks of Passenger Pigeons once darkened the skies; there were so many birds that they sometimes broke branches off trees when they landed. Now they are extinct. The abundance of other species that depend upon large tracts of mature forest reflect the status of our forests. These species would include Red-shouldered Hawks, Scarlet Tanagers, Ovenbirds, and Black-throated Blue Warblers. The densely forested northwestern parts of the county are particularly suitable for these species, but they may spread back into other areas if we allow our forests to mature again.

Some birds reach their northern breeding limit in the vicinity of Lanark County.[36] They include the Tufted Titmouse, Blue-winged Warbler, Cerulean Warbler, Louisiana Waterthrush, Yellow-breasted Chat, and Orchard Oriole. There is also a possible breeding record of a Carolina Wren near Carleton Place, and Blue-gray Gnatcatchers are now recorded from both Big Rideau Lake and Ottawa. Some more widespread birds appear to have provincially significant centres of abundance on the Frontenac Axis. Examples might include the Yellow-throated Vireo, Golden-winged Warbler, Cerulean Warbler and Prairie Warbler. Other widespread species appear particularly common on the Frontenac Axis; Pine Warblers and Whip-poor-wills have distributions which appear to follow the rocky open forests, and Hooded Mergansers apparently use the many small ponds of the axis for breeding. Some birds extend southward along the axis; these include the Swainson's Thrush, Raven, and Common Loon; we may think of these as ambassadors of Algonquin Park. There is even a remarkable breeding record of a true boreal forest species, the Three-toed Woodpecker, southwest of Big Rideau Lake.

With clearing for agriculture, prairie-like conditions have been created. Birds of open areas have likely increased in abundance as a result. Bobolinks and Eastern Meadowlarks would be two examples that one often sees along the edges of fields. Upland Sandpipers would be another example; they breed in pastures here, and in fact reach their densest concentration in southeastern Ontario. The Loggerhead Shrike is another species which may have benefited at first when fields replaced forests. First recorded in Ontario in 1860, it reached its peak abundance in the mid-1900s when it occurred as far north as Thunder Bay and North Bay. In spite of humans

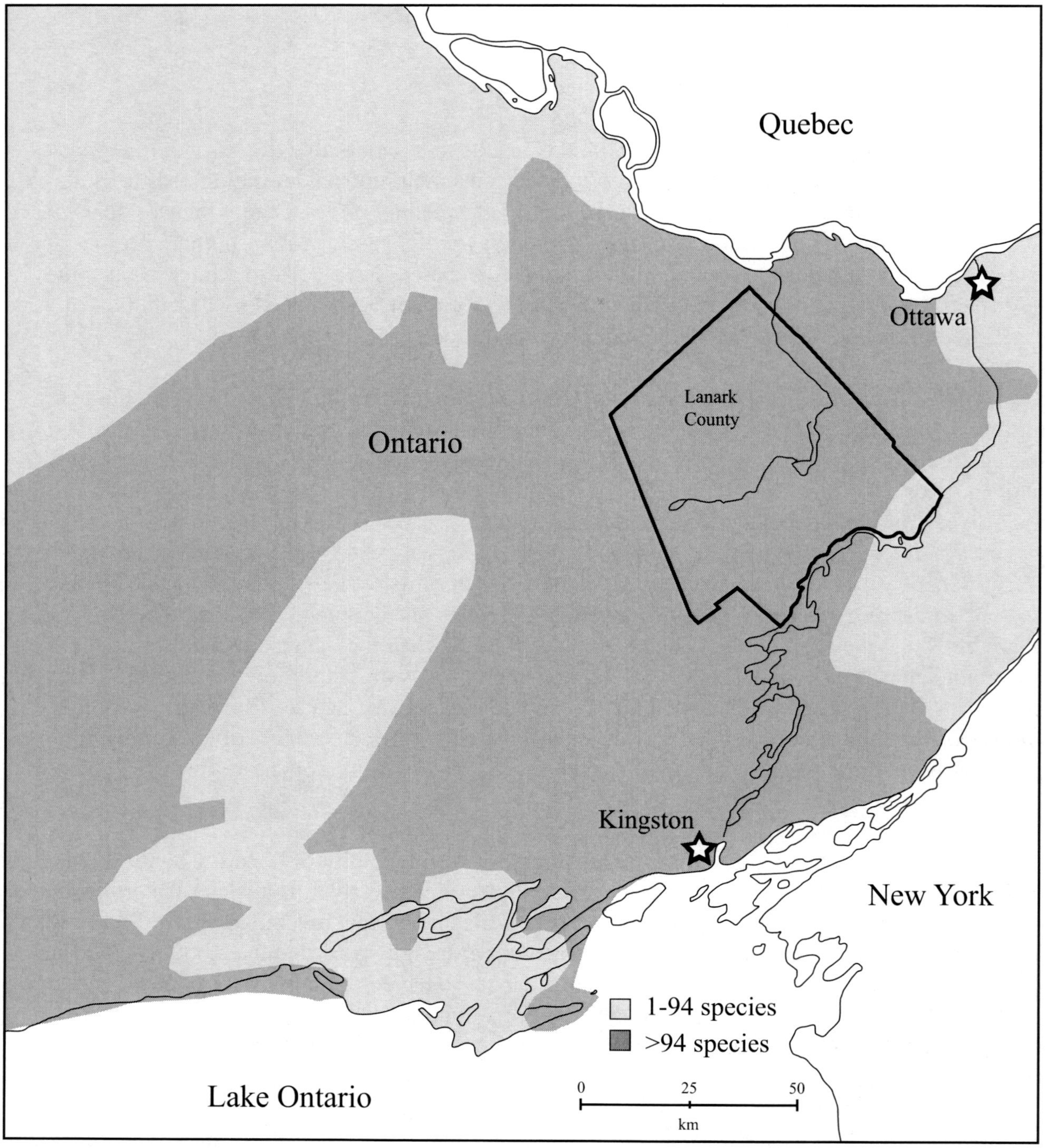

Map 11. Density of breeding birds in eastern Ontario.

creating more open conditions, the shrike is now on the endangered species list for Ontario because its populations have been in steady decline. Between 1991 and 2006, the population in the province as a whole ranged from only 18 to 55 breeding pairs. Some of them breed in southern Lanark County.

The extensive areas of wetlands also support many kinds of water birds. Great Blue Herons are a common sight on shorelines, and if you are lucky, you may see a heronry where the birds nest in large colonies. Sometimes Ospreys take over a heron nest and build their own nest on top. Ospreys can be seen hunting over our larger lakes. If you are very fortunate, you may see them dive into the water and emerge with a fish. On some of our lakes one can still hear and see Common Loons as well. These birds are indicators of the health of our wild areas, and appear to be in decline as the number of cottages and motorboats grows each year. Bald Eagles were once common in the Rideau Lake area — in 1952 one observer reported 15 nests within a 32 km radius of his cottage. While the species has declined significantly since that time as a result of pesticide use, habitat destruction and shooting, it is now confirmed to breed west and south of Lanark County and may possibly breed within it.

To learn more about the birds of Lanark County, you need a good field guide and binoculars. You might also consider going on an excursion with a local natural history society. Birds move quickly, and it helps to have an expert along to assist with identifying them.

Plants

The dominant plants in Lanark County are trees, and of these, there are only about a dozen common species. Altogether, though, there are more than one thousand taxa (species, subspecies, varieties, hybrids) of smaller plants in the county. These include about ten kinds of carnivorous plants such as round-leaf sundew and pitcher plants. There are about a dozen kinds of orchids, including the rose pogonia and calopogon that grow in peat bogs, the showy lady's slipper that grows in fens, and the showy orchis that grows in rich deciduous forests. There are even parasitic plants such as Indian pipe and beech drops, both of which no longer have green leaves, but rather feed off tree roots. Other species such as spring beauty and wild leek produce leaves for only a few weeks each spring and then vanish for the rest of the year. walking fern is found in a few areas with marble bedrock. The list of species would be longer still if kinds of mosses, lichens and algae were included. It is likely that new species will be found as botanists continue to explore the less known parts of the county. The most common plant is probably sugar maple (at least if we were to tally the weight of each species in the county). Among the rarest are a small species of primrose found in fens near Napier Lake, and a small spike moss found near Playfairville Rapids. A total of 375 native plants — 242 rare and 133 sparse — are considered significant in Lanark County.[37]

The kind of plants in a particular area depends largely upon the soil, the moisture supply, and human history. Most well-drained areas will have forests and, because of the shade produced by trees, most forest plants are either shade tolerant (as in many woodland ferns) or else they grow mainly in the spring when there are no leaves on the trees (as with trout lily and blue cohosh). Wetland areas have different plants from uplands, and the kind of plants will depend upon whether the wetland is a swamp, marsh, bog or fen. These kinds of wetlands are discussed further in the next section. Dry rocky open areas on the Precambrian shield such as those north of Christie Lake and on Pakenham Mountain create prairie-like rock barrens. Rocky open areas on limestone create prairie-like plant communities known as alvars. The list of special places at the end of this book provides better descriptions of the different kinds of habitats in the county, and some of the unusual plant species found there.

One kind of habitat — abandoned fields, pastures and road edges has been created by humans. The plants here are very different from the rest of the county in that many of them were only recently introduced from Europe. Since most of Lanark County had species adapted to forests, most of the plants in this county were unable to tolerate the open sunny conditions created by settlers clearing the forests. This created a vacancy which was occupied all too soon by weeds that invaded with seeds and livestock from Europe. Many of the beautiful wild flowers that bloom each July through September in ditches and fields are of European origin. These include species such as ox-eye daisy, orange hawkweed and evening primrose. Although they are beautiful to look at, they also are constant reminders of just how much humans have changed the environment of this area. Many of the species of aster and goldenrod found in these fields are, however, native; prior to the arrival of Europeans, such species probably survived in the dry open rock barrens and alvars of the county. Others likely invaded from areas farther west along roads and railways.

Wetlands

Wetlands are some of the most biologically productive places on the planet. With their natural irrigation, they produce both oxygen to breathe, and wildlife to enjoy. Fish, waterfowl, and most fur-bearing animals require wetlands to survive. Map 12 shows that some areas of Lanark County, particularly Drummond and Beckwith townships, were nearly half wetland when the first settlers arrived! At that time, when no one thought much about the possible disappearance of wildlife, and when the importance of wetlands was little understood, they were drained as fast as possible. As a consequence, large areas of wetland have been lost. Map 13 shows that in Ramsay Township, roughly two thirds of the wetlands have been drained. But in Beckwith Township, and in the northwest townships, less than twenty percent of the wetlands have been drained. This means that Lanark County is still relatively rich in wetlands compared to many other parts of southern Ontario. Yet many of the areas of clay and muck soils shown in Map 6 were once wetland. These areas are now largely farmland, but the presence of drainage ditches is a constant reminder that it is only by continually removing the water that wetlands are prevented from returning. Drainage ditches in these areas can be distinguished from streams because natural streams meander, whereas drainage ditches follow straight lines. We may decide some day that some areas of marginal farmland should be allowed to return to their natural state. A wetland, of course, is a product of water. Areas that are flooded long enough to form distinctive soils and have distinctive plants that tolerate flooding, are called wetlands. We have four main kinds of wetlands:

(l) Swamps are wetlands with organic soils and standing trees. Most wet depressions here have ash, elm and red maple in them. Along major water courses are some large silver maple swamps. These are best seen by canoe, and one fine example is part of the Mississippi River floodplain between Appleton and Almonte.

The largest turtles in Lanark County are snapping turtles. They are slow moving, feed largely upon carrion, and rarely leave the water except to nest. Snapping turtles are most often seen on the edges of roads in June when the females are looking for warm locations to bury up to 20 round eggs. The young turtles emerge in September and have to find their way to water. Since the shell is small in proportion to the fat body, these turtles cannot completely fit within their shell. When in the water they will normally flee from humans, but they snap vigorously when on land and frightened. If left alone, they are harmless.

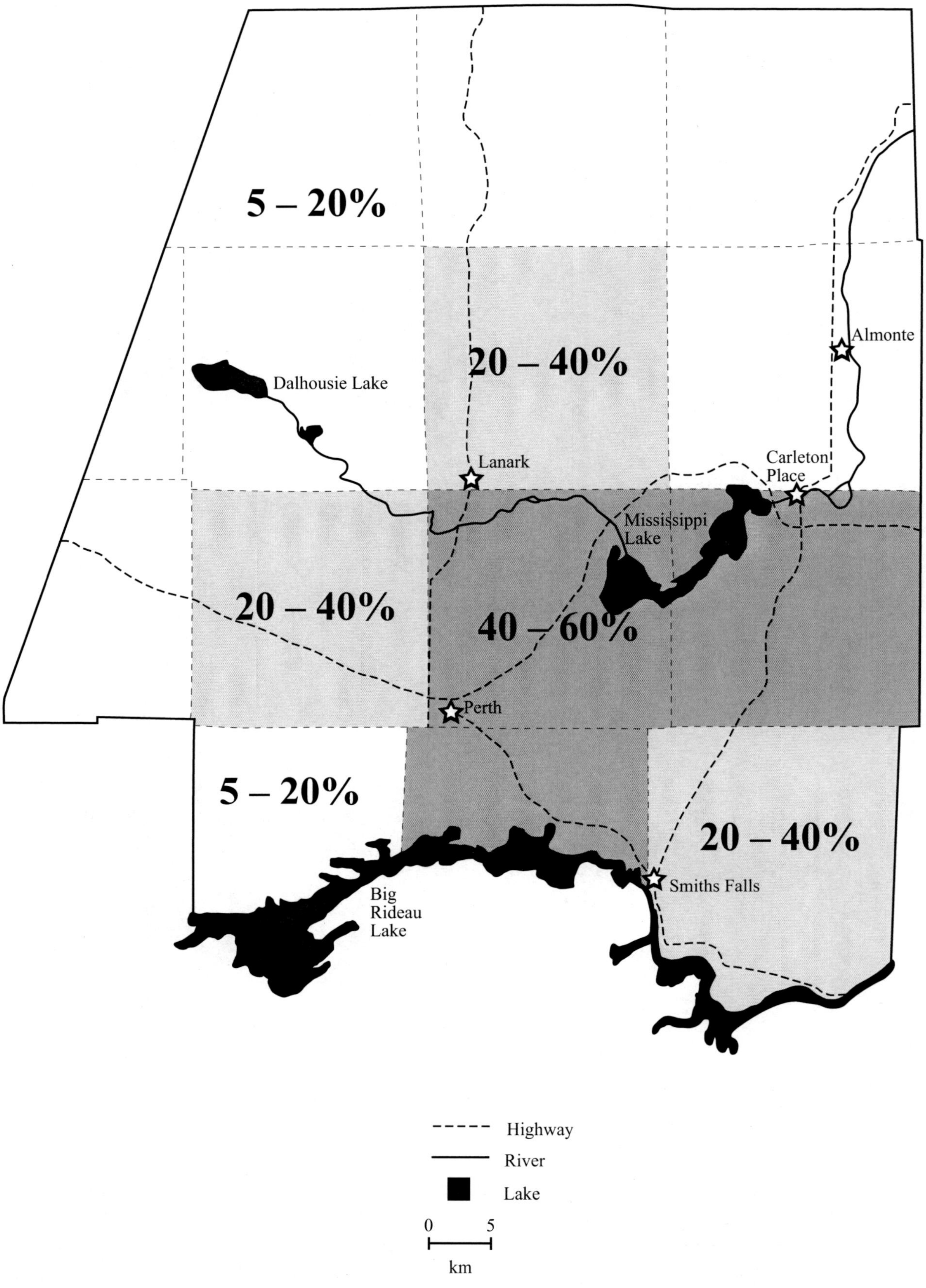

Map 12. Percentage of Lanark County originally wetland.

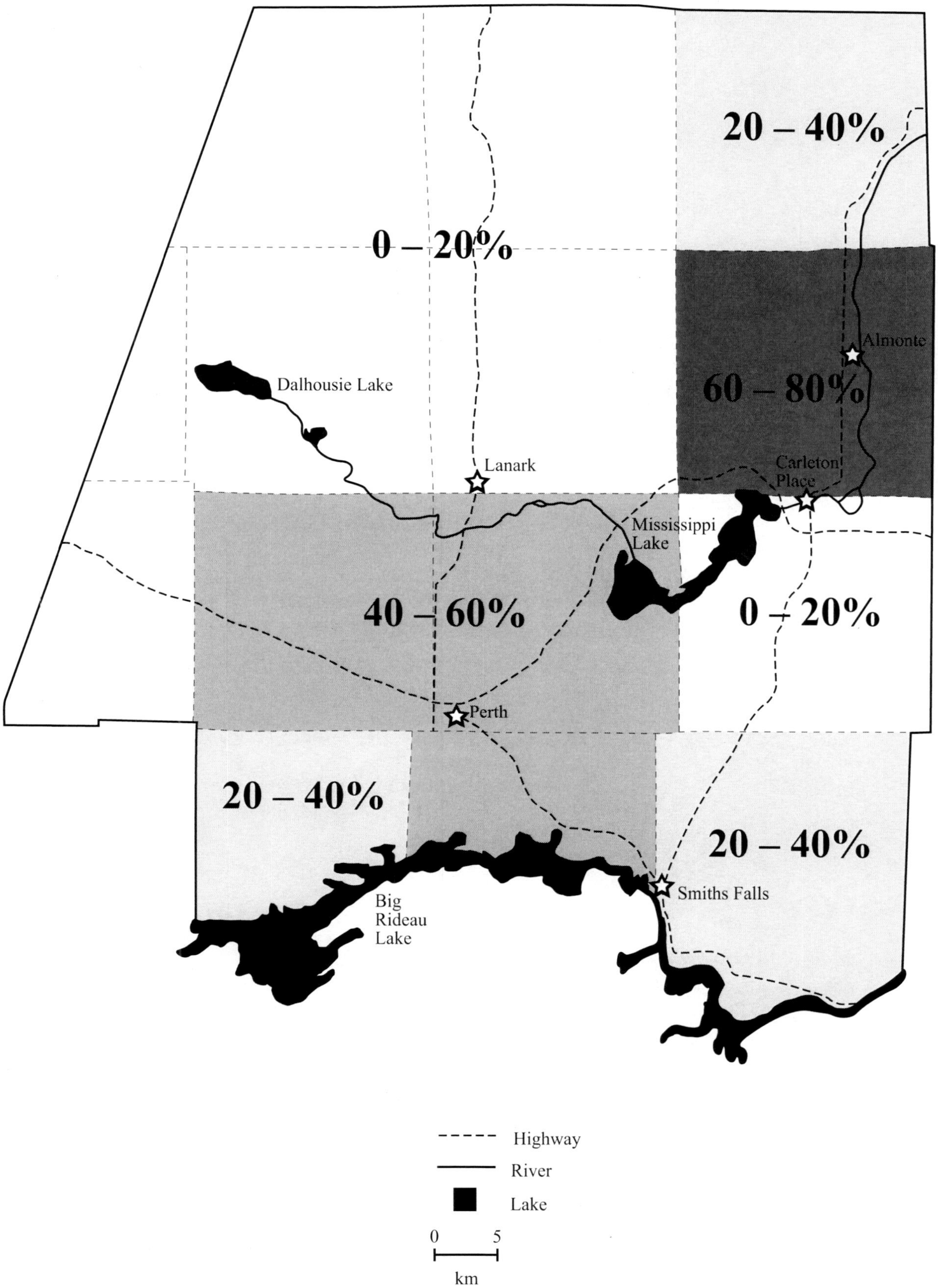

Map 13. Losses of original wetland from *c.* 1800 to 1982.

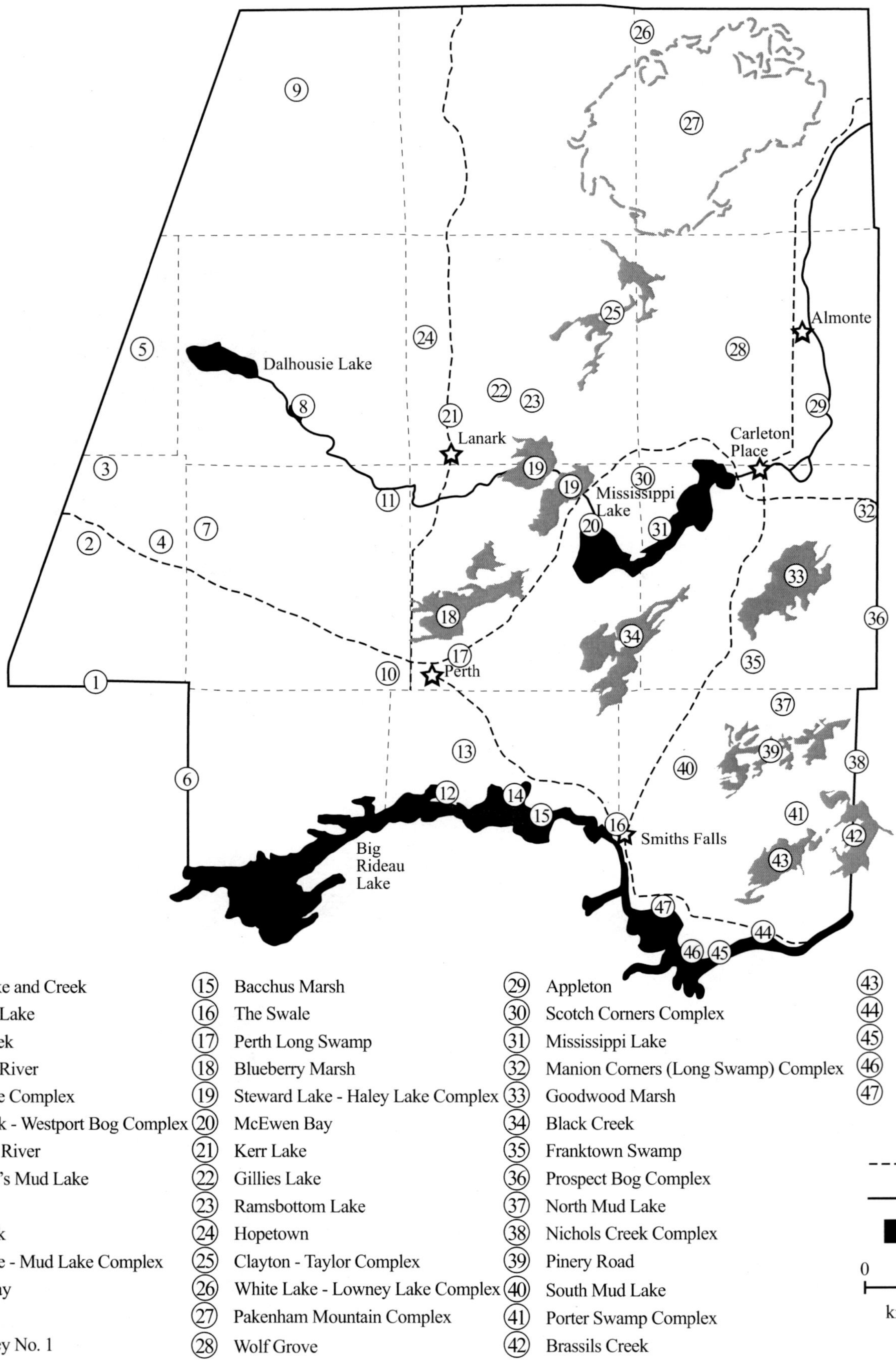

1. Crosby Lake and Creek
2. Little Mud Lake
3. Bolton Creek
4. Upper Fall River
5. Stump Lake Complex
6. Black Creek - Westport Bog Complex
7. Lower Fall River
8. McCulloch's Mud Lake
9. Joe's Lake
10. Grant Creek
11. Playfairville - Mud Lake Complex
12. McLean Bay
13. Tay Marsh
14. Port Elmsley No. 1
15. Bacchus Marsh
16. The Swale
17. Perth Long Swamp
18. Blueberry Marsh
19. Steward Lake - Haley Lake Complex
20. McEwen Bay
21. Kerr Lake
22. Gillies Lake
23. Ramsbottom Lake
24. Hopetown
25. Clayton - Taylor Complex
26. White Lake - Lowney Lake Complex
27. Pakenham Mountain Complex
28. Wolf Grove
29. Appleton
30. Scotch Corners Complex
31. Mississippi Lake
32. Manion Corners (Long Swamp) Complex
33. Goodwood Marsh
34. Black Creek
35. Franktown Swamp
36. Prospect Bog Complex
37. North Mud Lake
38. Nichols Creek Complex
39. Pinery Road
40. South Mud Lake
41. Porter Swamp Complex
42. Brassils Creek
43. South Montague
44. Merrickville Marsh
45. Rideau River Part 4
46. Rideau River Part 3
47. Rideau River Part 1

Highway
River
Lake

0 5
km

Map 14. Provincially significant wetlands (shaded are >1,000 ha) in Lanark County. Within the perimeter shown for wetland complex 27, myriad, interconnected small wetlands occur.

(2) Marshes are areas which are flooded so much that trees cannot survive in them. They are therefore wetter than swamps. Marshes are often covered by cat-tails, bulrushes, sedges and reeds. There are many small marshes scattered around the county; larger marshes tend to be associated with bays of lakes such as Mississippi Lake and Big Rideau Lake. When beavers flood swamps, they can drown the trees and turn them into marshes.

(3) Fens are an uncommon wetland type in this part of the world, and most people may have never seen one. Fens have organic soils and slowly moving water which carries nutrients through the soil. This distinctive water and nutrient situation supports unusual vegetation types with many rare plant species. Fens are a good place to find orchids. There are few fens in Lanark County, but one of the most accessible is the Purdon Orchid Bog. (Although it is called a bog, it is really a fen.) It has a board walk through it so you can explore it with the family and still keep your feet dry. Another large fen is found along the shores of White Lake.

(4) Bogs form where water levels prevent decomposition and a thick layer of peat accumulates. Mosses such as *Sphagnum* moss are a major component of this peat. This wet, organic, acidic environment is so difficult for plants to tolerate that only a small group of distinctive species can occur. Many are small ericaceous shrubs with evergreen leaves and colourful flowers. Because soil nutrients are so scarce, some of the plants in bogs catch insects as a source of vital nutrients; both pitcher plants and sundews can be found in the bogs of Lanark County. We have no big bogs on the scale of Mer Bleu or Alfred Bog, both of which are east of Ottawa. Mer Bleu is well worth a visit, being only an hour away. There is a board walk to explore the bog without getting wet feet. The Purdon Orchid Bog, mentioned above, is more a fen than a true bog because it is in an area with alkaline soils, and there is some movement of water through it. The Perth Bog is of particular interest because samples of ancient peat have been taken from this bog. By examining ancient fragments of wood and pollen grains, scientists have been able to learn more about events after the last ice age. The biggest bog in the county appears to be one south of Maberly.

In order to help identify and protect our remaining important wetlands, the Ontario Ministry of Natural Resources has a program of wetland evaluation. Each wetland is visited by a professional biologist, and is then given a report card which assesses its importance in four categories (biological, social, hydrological, special features). In each category a wetland may receive up to 250 points. Wetlands with a total of 600 points or 200 points in the biological or special feature categories are considered provincially significant. There are now provincial planning guidelines to protect provincially significant wetlands. Landowners with such wetlands also receive a subsidy on their property taxes as an incentive to keep them in their natural state. Map 14 shows the provincially significant wetlands designated to date. Evaluation is an ongoing process, so new sites may be added in the future.

Beaver Ponds

Beaver ponds are a distinctive kind of wetland. It is hard to drive for more than half an hour in Lanark County without seeing one somewhere. As everyone knows, beavers create wetlands by damming small streams. What may be less appreciated is that beavers are often repairing damage caused by our great-grandparents — by blocking old drainage ditches, beavers have probably restored thousands of acres of wetlands in Lanark County over the past few

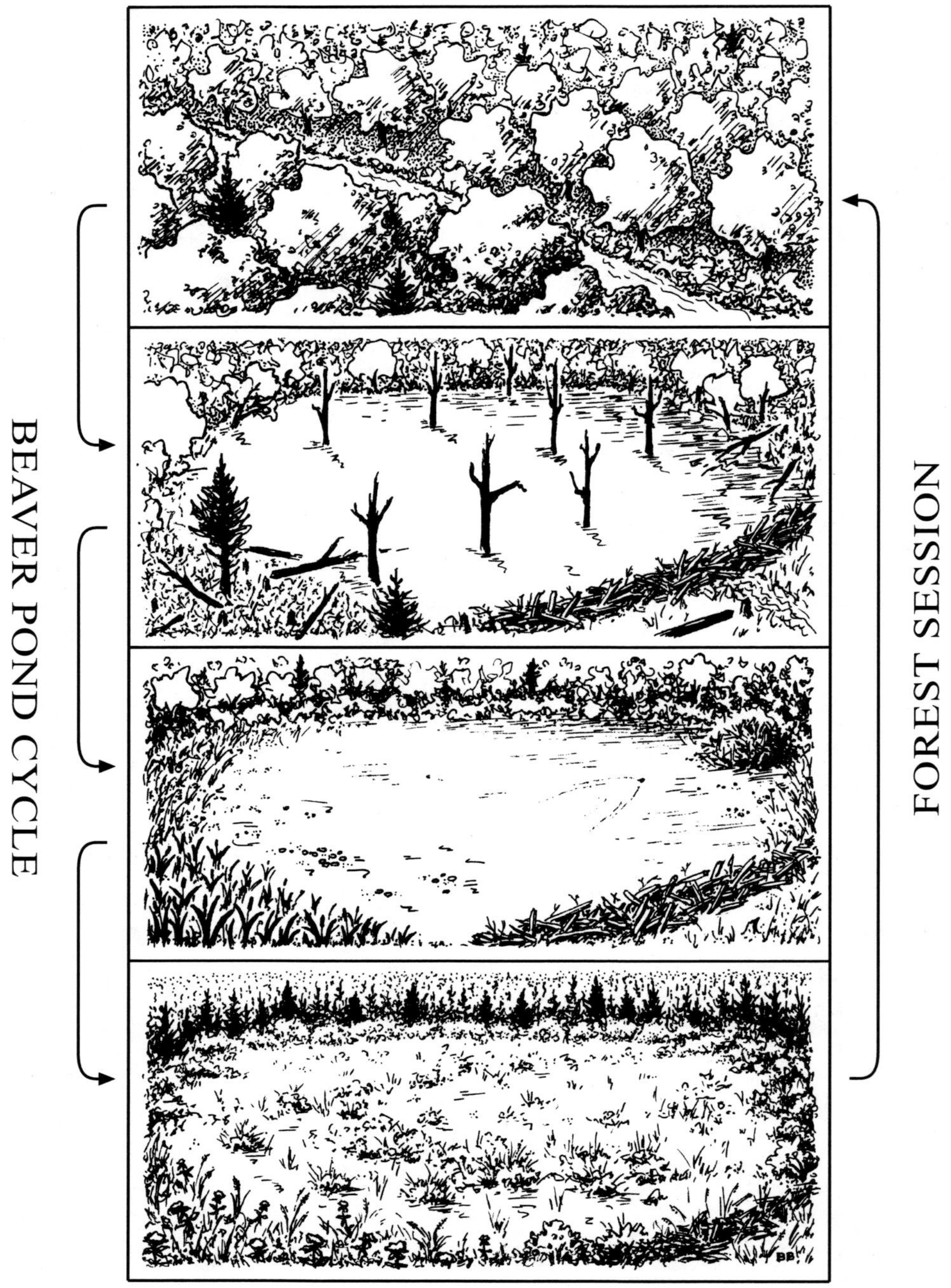

Figure 2. Beavers can change valleys with small streams (1) into a wide array of other wetland habitats including ponds (2), marshes (3) and wet meadows (4). If beavers remain absent for enough time, forests will regrow (1), and the cycle can begin again (illustration provided by Betsy Brigham).

decades. Our great-grandparents would probably be mortified to see their hard dug ditches filling in and fields being reclaimed by marshes. It is hard to believe that it has taken only a hundred years for wetlands to become as valuable as farmland.

Beavers are particularly interesting because they not only create wetlands, they switch one kind of wetland to another by changing water levels. High water periods kill woody plants and remove areas of swamp forest. Low water periods create mud flats on which marsh plants can establish. To understand such wetlands, one has to think about how they change over time. Dams are often temporary. Once the local food supply of species such as poplar is exhausted, the beavers may starve or leave. Once the beaver dams are no longer maintained, water levels fall and mud flats are exposed. There are vast numbers of buried seeds in the mud (tens of thousands of seeds per square metre is common), and on drying mud flats there is a burst of germination of plants such as bulrush, cat-tail, bur-reed, smartweed, sedges and rushes. Such periodic low water periods are necessary for many wetland plants to germinate and re-establish. Since many wetland plants are important food for waterfowl, management groups such as Ducks Unlimited deliberately create temporary low water periods to improve wetlands for wildlife.

If water levels remain low for a number of decades, then woody species begin to grow again. In Lanark County, willows appear to be the first woody plant to invade, but most of these areas would probably return to ash and red maple forest.

As Figure 2 shows, beaver ponds therefore naturally cycle among a number of vegetation types (swamp forest, shrub thickets, open water, mud flat, marsh, wet meadow) depending upon recent water levels. Any single wet depression can cycle through all of them given sufficient time. It may take only a few years to switch among some vegetation types (say from open water to wet meadow) but others would obviously take much longer (probably at least a century to go from wet meadow to swamp forest). A fine beaver meadow occurs in a valley on the south side of Highway 7 about 10 km west of the Carleton Place junction. Most of the different stages of the beaver pond cycles can be seen in a short drive along the concession roads north of Highway 7. See if you can find them.

The Frontenac Axis: A Biological Highway

The western edge of Lanark County is part of a narrow extension of the Canadian Shield which links the Madawaska and Algonquin Highlands with the Appalachian Mountains far to the south (Map 15). This extension is called the Frontenac Axis. It can be seen distinctly along Highway 401, where, between Kingston and Brockville, the fertile agricultural lands typical of southern Ontario are replaced by rock outcrops reminiscent of Algonquin Park. This provides a unique combination of environmental conditions — the rocky shallow soils typical of Canada's north, but a distinctly southern climate. Trees such as pitch pine occur only on the southern portions of the Frontenac Axis. The Frontenac Axis also has some of the highest numbers of breeding birds of any region of Ontario, with southern birds such as the Blue-gray Gnatcatcher and the Cerulean Warbler nesting as far north as the Big Rideau Lake area. This is also prime habitat for reptiles such as the Gray Ratsnake, our only tree-climbing snake. The headwaters of the Mississippi River lie in the axis. As climates continue to change, we may think of the Frontenac Axis as a giant biological highway linking northern Ontario with the Appalachian

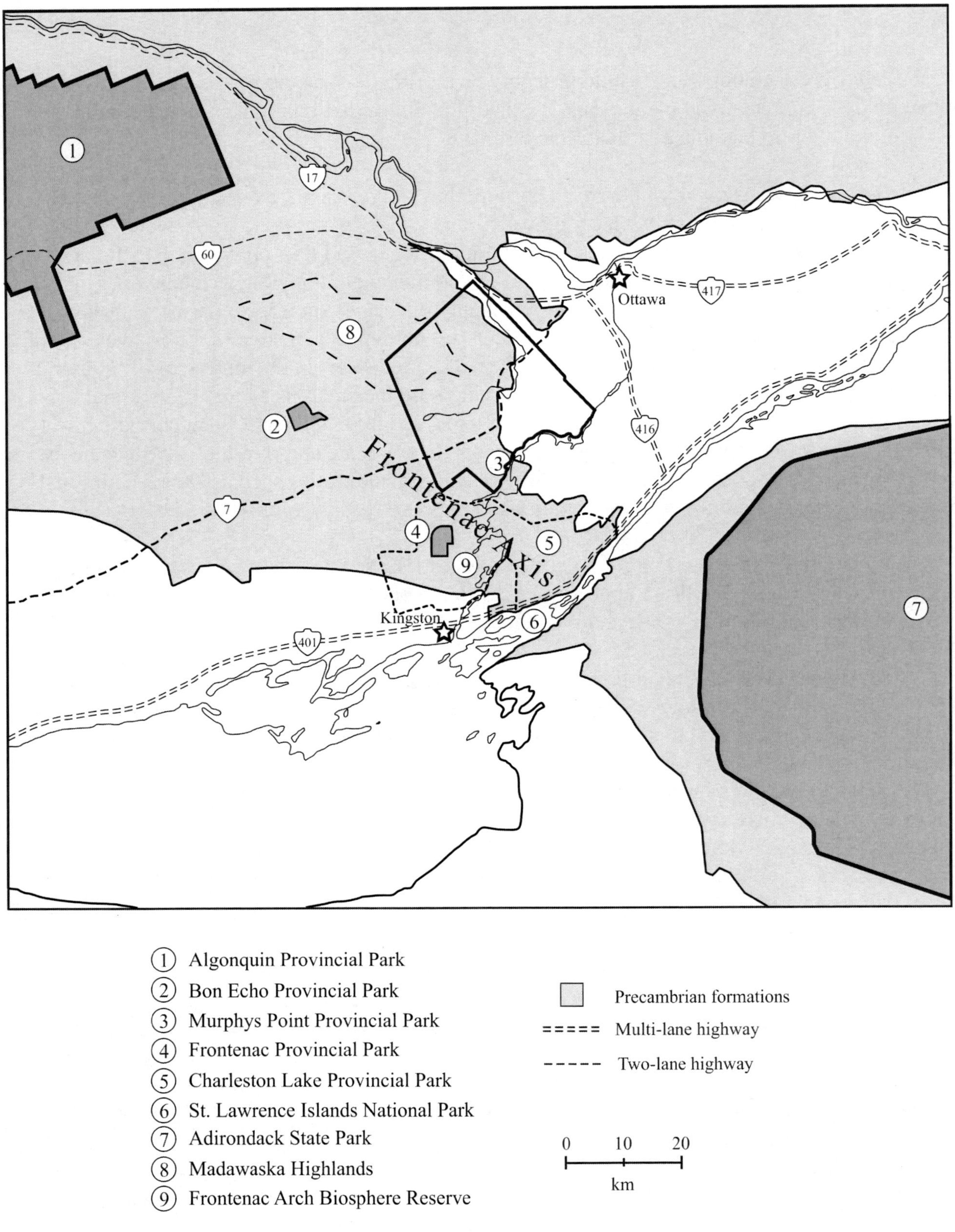

Map 15. The Frontenac Axis, which links Algonquin Park to the Adirondacks, passes along the edge of Lanark County.

Mountains. Lanark County lies on the edge of this highway; in fact, the Precambrian rocks in Lanark County are actually part of the Frontenac Axis. A portion of the axis has been designated a biosphere reserve under the Man and the Biosphere Programme of the United Nations Educational, Scientific and Cultural Organization (UNESCO).

Long Lake and Flower Round Lake: Our Unofficial County Archives

All our lakes formed at the end of the ice age. Each began with a bottom which had been scoured clean by moving ice and running water. Most received a layer of clay from the period when they were covered by glacial lakes and the Champlain Sea. Since then, they have slowly accumulated debris from the surrounding landscape. Clay from eroding stream banks. Pollen from trees. Ash from forest fires. This debris accumulates in layers in the bottom of the lake, providing a steady record of the environment around in the lake. If a core of mud is taken from the bottom of the lake, it can be examined to reconstruct the events in the lake's history for the last 10,000 years.

Two Lanark County lakes have been studied in detail by scientists at Queen's University: Long Lake (North Burgess Township) and Flower Round Lake (Lavant Township).[38] Both tell a similar story. Both began about 9,500 years ago when the ice melted back from Lanark County. For several thousand years after the ice age, there was rapid infilling of the lakes from erosion around the lakes. As plants reinvaded, the soil was stabilised; rates of erosion decreased significantly around 7,500 years ago. The early forests were pine and spruce forests, similar to those currently found much farther north, but deciduous forest trees such as maple, oak and beech migrated into the area about 7,500 years ago. We can deduce this from the rapid appearance of their pollen in the lake sediments at this point.

A major event occurred about 5,000 years ago when hemlock trees apparently died *en masse.* This die off of hemlock occurred throughout eastern North America. Since there were no scientists keeping records then, the cause of hemlock decline is still unknown. It seems likely that a disease spread rapidly, much the way that Dutch elm disease removed elm trees from our landscape in the 1960s. This disruption of the forest led more nutrients to erode into lakes, briefly increasing their productivity.

The most dramatic changes occurred in the past century as forests were cleared, roads were constructed and weeds invaded. Rates of sedimentation increase and pollen from weeds becomes abundant in lake sediment. In fact, ragweed, which causes so much human suffering each summer, invaded with the settlers. Its sudden appearance in lake sediment is so dramatic that ragweed pollen is now used (in part) to determine which portions of sediment cores date from periods after human settlement.

One of the reasons for studying such lake cores is that they allow us to compare past events to present day conditions. Much of what I have written here about the history of forests has come from the study of such cores across North America. One of the more sobering results of these detailed studies is that none of the past events — changes in climate, forest fires, hemlock die back — has caused as dramatic a change in our landscape as humans have.

Spring peepers are tiny tree frogs whose distinct "peep-peeeep-peep" can be heard in marshes and ponds from late April through to June. Easily audible from cars with open windows, these frogs are a true sign of spring, for they begin breeding even while ice still remains along the edges of ponds.

The deep cow-like drone of the bullfrog is mostly heard in the distance during hot summer nights. Tadpoles of this species may take several years to mature into frogs. Unlike the tiny spring peepers, which breed in temporary ponds, bullfrogs are large and require deeper permanent water.

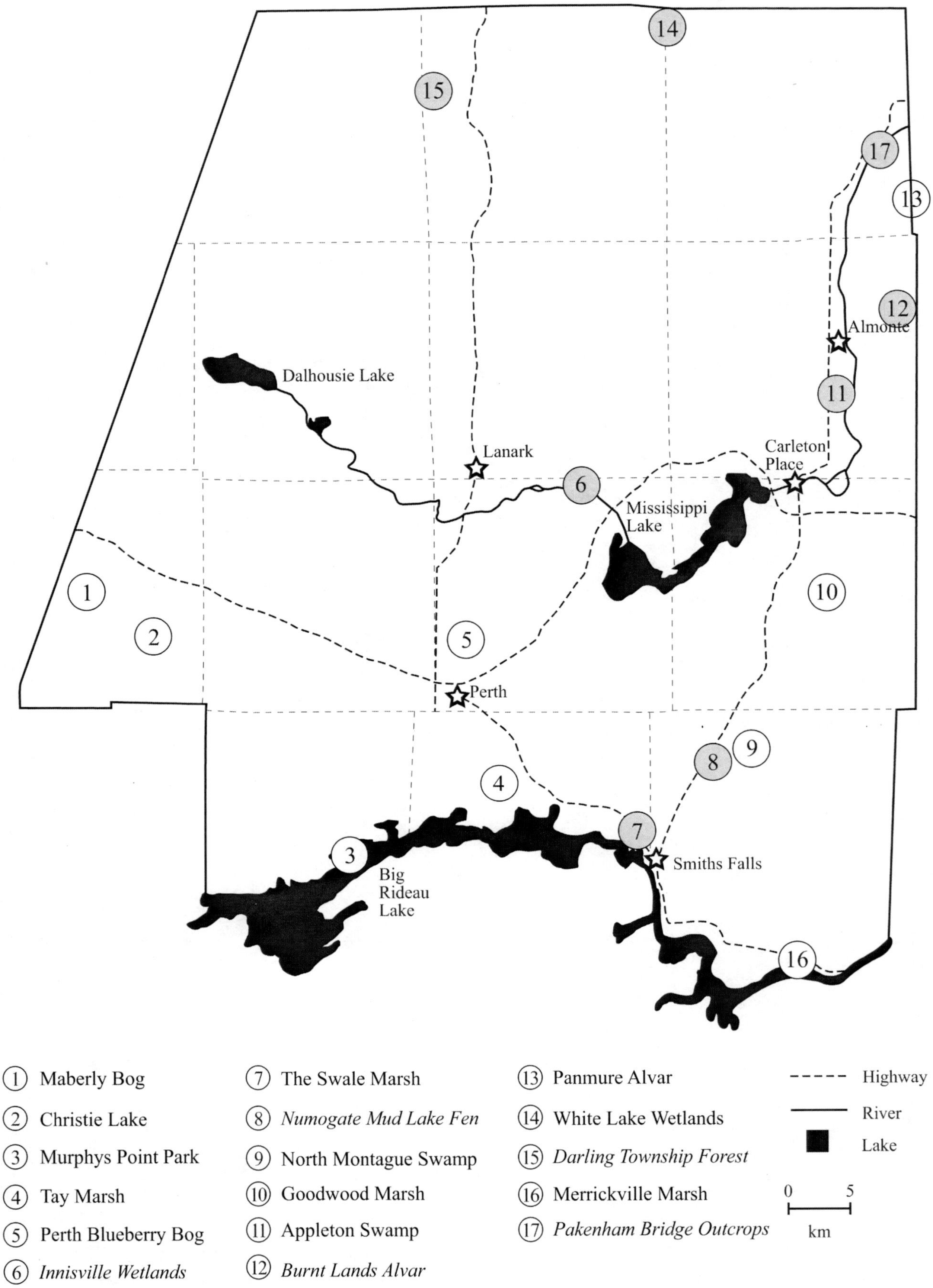

Map 16. Provincially (shaded) and regionally significant Areas of Natural and Scientific Interest in Lanark County. (italics = confirmed by the Ontario Ministry of Natural Resources and Forestry, plain text = identified as candidate ANSIs by OMNRF).

Special Places

Although it is close to the capital city of Canada, there are still large gaps in our knowledge of Lanark County. We do not have a comprehensive list of important natural areas and natural features, and most of those that we know about are not adequately protected. We need to find the important natural areas and natural features of the county, and we need some way to protect these areas for the benefit of ourselves and future generations. One positive step in this direction is the wetland evaluation system. Another is the program of the Ministry of Natural Resources to identify ANSIs — Areas of Natural and Scientific Interest. Map 16 shows the ANSIs identified to date. More remain to be found. It should be noted that sites that are designated ANSIs can remain in private ownership and currently have only limited legal protection — the idea is to let citizens know that these areas are important, and hope that this moral persuasion will help to protect them. Sometimes it works, sometimes it does not. Please note that many are on private land — the fact that the areas are ANSIs does not guarantee that you can visit them.

Map 17 shows some examples of interesting natural features in Lanark County. Some of these may be documented as ANSIs, such as the Burnt Lands Alvar, or protected as public conservation land such as the Purdon Conservation Area. Others are just places I like, such as the Lavant/Darling Spillway. Some may be on private land and can be appreciated only as you drive by. The fact that they are here in this book does not mean they are open to the public, or that I am responsible if you trespass and get eaten by a bear, drowned, or shot by an irate landowner. If I know the land is public I say so, but for some I do not know the ownership, and even if I did today, it could change by tomorrow. Moreover, if you are not used to maps, it is easy to get confused over boundaries and wander off publicly owned land. So be careful and err on the side of politeness and caution.

One other warning is necessary. The intent of this section is to provide a selection of sites which may interest a wide range of readers including (1) arm chair biologists wanting to browse on a rainy day, (2) families looking for an educational and entertaining outing, and (3) avid field naturalists wanting to further explore unusual areas. The list below is by no means exhaustive or complete, nor are sites listed in order of importance. Since the initial publication of this book, I have prepared a conservation priority list for the county, and it is provided in Appendix A.

Christie Lake Fire Barrens (Area 1)

To the southwest of Lanark County there are extensive areas of open rock ridges with scattered trees. The few trees are generally those that are resistant to fire: red oak, bur oak and large-toothed poplar. There are open grassy areas with shrubs such as blueberries, junipers and sumac. Unusual plants typical of prairie habitats farther south occur too, including, in one case, a species of prickly pear cactus! There are also unusual reptiles, including rare species such as Five-lined Skinks (Ontario's only lizard) and spotted turtles. Rare birds such as Prairie Warblers are also found. Such barrens are particularly well-developed in the Kaladar area. How did these prairie-like habitats develop? Rather little is known about them at present, so it is a difficult question to answer. One view is that these open rocky barrens are a natural part of Ontario's vegetation, and that they are merely an extension of prairie areas farther south. It was once

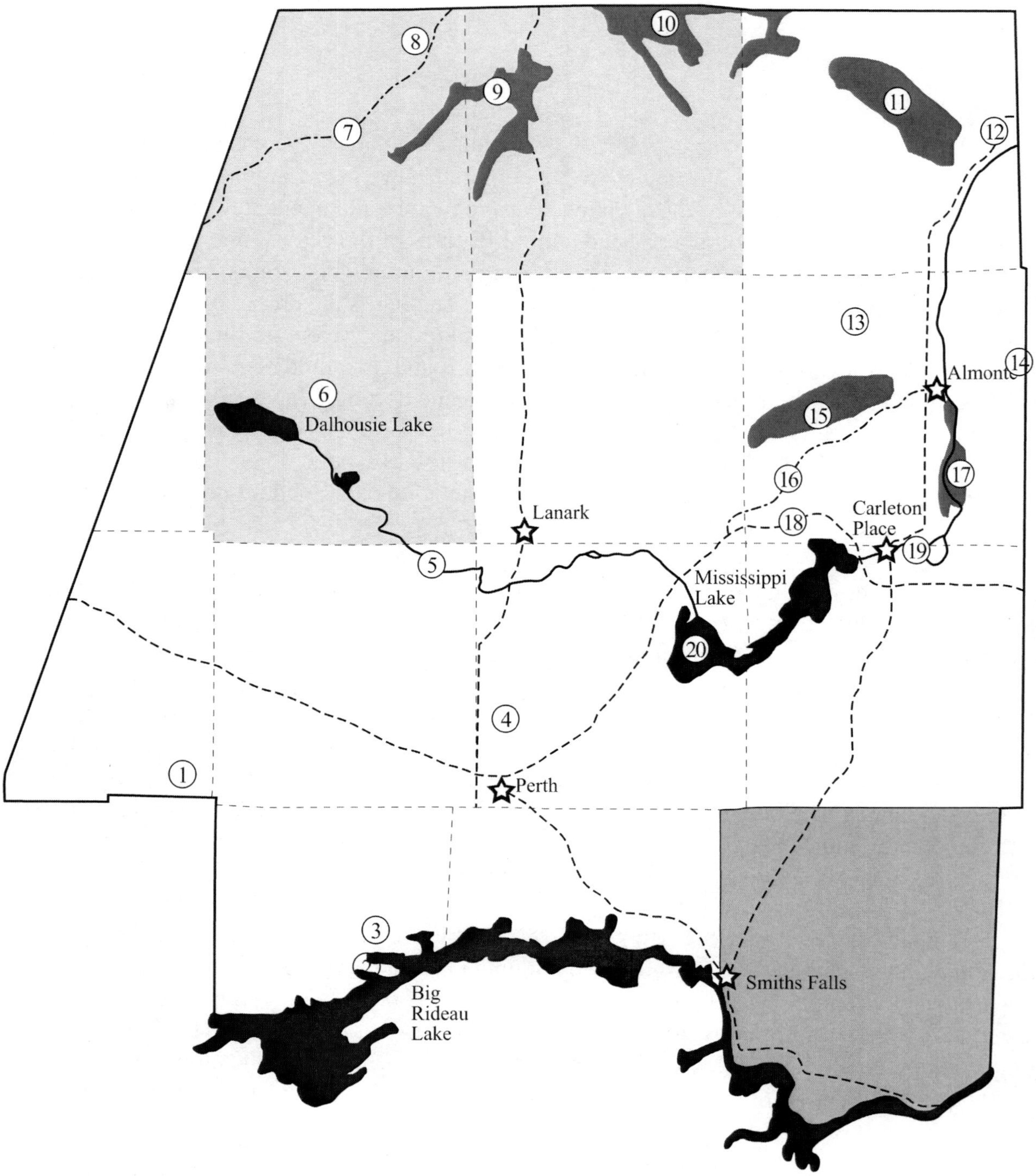

1. Christie Lake Barrens
2. Murphys Point Park
3. Long Lake
4. Perth Bog
5. Playfairville Rapids
6. Purdon Conversation Area
7. Kingston-Pembroke Railway
8. Flower Round Lake
9. Lavant / Darling Spillway
10. White Lake
11. Pakenham Mountain
12. Pakenham Bridge
13. Mill of Kintail
14. Burnt Lands Alvar
15. Wolf Grove
16. Old Perth Road
17. Appleton Silver Maple Swamp
18. The Edge of the Shield
19. Carleton Place Hackberry Stand
20. Mississippi Lake

Northern Forests
Shrike Nesting Area

Highway
River
Lake

0 5
km

Map 17. Special places in Lanark County.

thought that Ontario had very little natural prairie, but we now know that prairies and savannas came at least as far north as Rice Lake. Another explanation for these habitats is that they are largely a result of wild fire after settlement and logging. Perhaps both factors are at work: natural barrens may have expanded after fires caused by humans.

Lanark County has some barrens in the extreme southwest where fires may have swept in from the Kaladar area. A particularly fine example of barrens occurs north of Christie Lake, and this has been identified as an Area of Natural and Scientific Interest. Smaller areas of barrens occur elsewhere, such as the north shore of White Lake. Five-lined Skinks have been found as far north in the county as Clyde Forks (Map 18) — a surprise for those who expect to find lizards only in deserts and tropical forest.

Murphys Point Provincial Park (Area 2)

Murphys Point is a fine place to introduce yourself to the Frontenac Axis, the tongue of Precambrian rock that links the Algonquin Highlands with the Appalachian Mountains. There are some fine deciduous forests with both parking and trails. You may wish to visit them in late May or early June when the spring flowers are at the peak of bloom. Spring is also the time to listen for southern birds that reach their northern limit here — birds such as the Cerulean Warbler and Blue-gray Gnatcatcher. Check whether a conducted walk is available from the park naturalists. Watch too for Gray Ratsnakes (Map 18), the spectacular tree-climbing snake that occurs on the Frontenac Axis. Murphys Point Park is too small to protect sustainable populations of such species; we will need much larger areas of the Frontenac Axis left in a natural state if they are to survive into the next decades.

The Silver Queen Mica Mine is found in Murphys Point Provincial Park. There is a trail to the mine, and exhibits. If a settler found it impossible to make a living by growing food on top of the ground, there was always the possibility of generating extra cash by selling what lay beneath it. Many small mines were established around the turn of the century. Mica is a mineral which forms large crystals which can be split into translucent layers. These were therefore sometimes used to make windows in stoves. As the use of electricity spread, there was increasing demand for mica to make insulators. Regular guided tours of the mine are offered by the Ministry of Natural Resources. Enquire at the park office.

Long Lake and Flower Round Lake (Areas 3 and 8)[39]

Long Lake (North Burgess Township) and Flower Round Lake (Lavant Township) share one key feature: they have been studied in detail to reconstruct their history over the past 10,000 years. As described earlier, sediment cores were taken from these lakes in 1988 and 1989, respectively, by a team from the Department of Biology at Queen's University. Study of the sediments and the remains of pollen and algae they contain has provided some of the best information that we have on the environmental history of Lanark County. When you view either of these lakes remember that their sediments are an unofficial archive of events in Lanark County. The chemistry of the water, the aquatic life and life in the surrounding forests has changed through time. ragweed, DDT, heavy metals, road salt, acid rain, and phosphates from dishwashers are now leaving their own unofficial record of our activities.

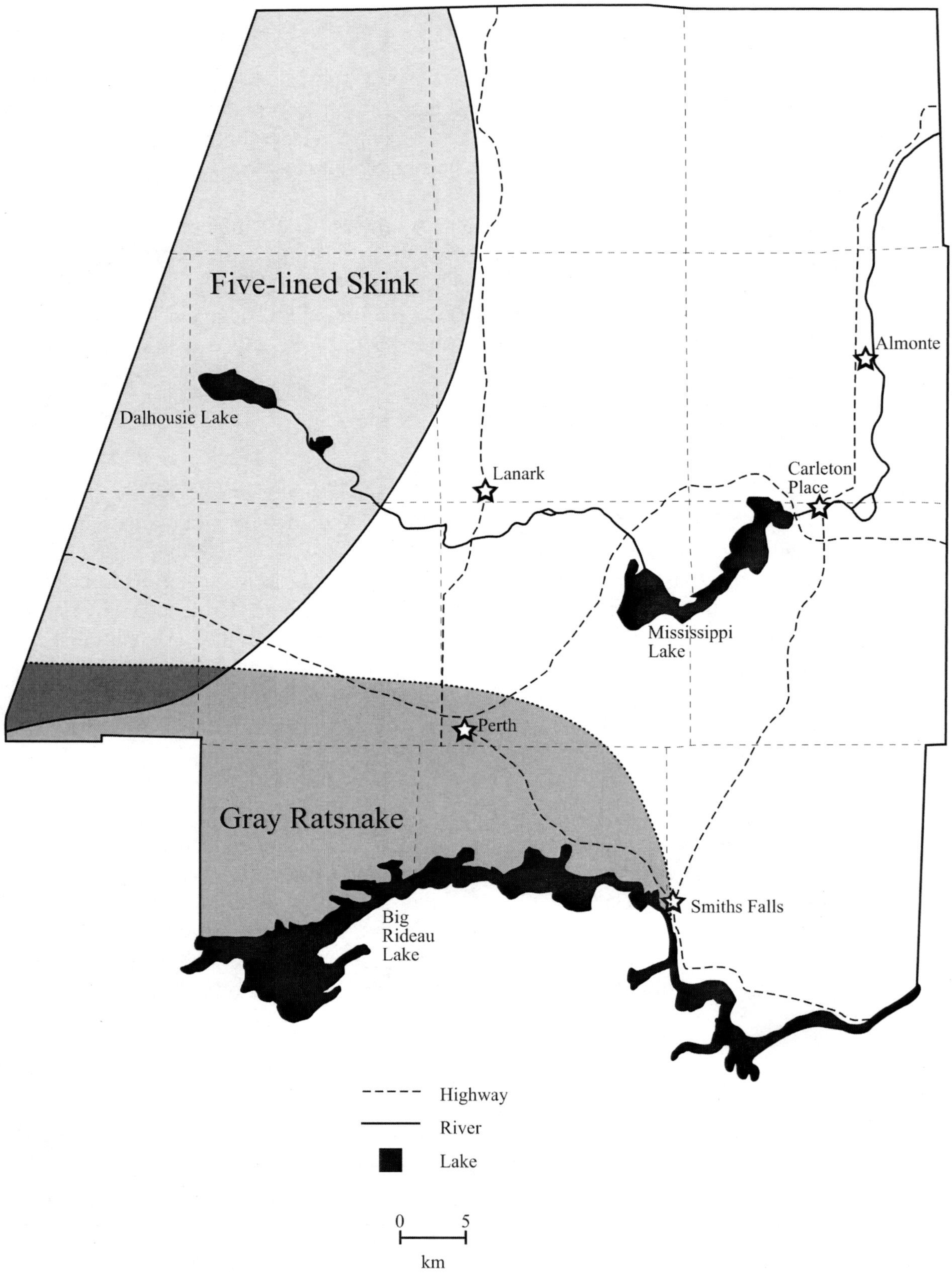

Map 18. Distribution of Five-lined Skinks and Gray Ratsnakes in Lanark County.

These lakes also remind us about the importance of scientific research. What we now know about environmental history does not just come from books, it comes from the labours of scientists at our universities who do this work. It is often assumed that professors only teach, but half our time is spent on research. If we did not do it, it would not get done.

The Perth Bog (Area 4)

The Perth Bog has developed in a depression where water has prevented the luxuriant plant growth from decaying. Over the years peat has accumulated to a depth of many feet. Underneath the peat is clay which was deposited under water at the end of the last ice age. Geologists are using information from the clay and the peat to reconstruct what happened to this part of Lanark County at the end of the ice age. How long was it under water? Was it fresh or salt water? What were the early forests like? Fragments of dead wood from deep in the peat inform us that the forests here once were largely spruce and hemlock — quite different from present day forests. By radiocarbon dating, it is possible to determine how old the wood is and how long ago these types of forests were here. There is much detective work remaining to be done, and most of the clues are still buried in the peat. This area is both a provincially significant wetland and an Area of Natural and Scientific Interest.

Playfairville Rapids (Area 5)

On the shorelines of these rapids one can find another unusual group of plant species. They include grass-of-parnassis (not a grass, but a white-flowered herb with glossy leaves not unlike plantain) and an unusual spike moss. Both of these plants are often found in wetland areas around the Great Lakes. Obviously something unusual happened to allow them to spread here, and the environment is apparently distinctive enough that they can survive only here. Perhaps it has something to do with seepage of calcareous water from the adjoining land. Although the land here is private, you can get a fine view of the shoreline by looking upstream from the bridge at Playfairville. You can also see rapids and an old mill.

Purdon Conservation Area (Area 6)

The Purdon Orchid Bog is not a bog — it is a fen. Fens are a rather unusual kind of wetland in Lanark County. The Purdon fen is composed of peat which has slowly accumulated on the shore of Purdon Lake. Unusual habitats often support unusual species. This one is noteworthy for a large colony (about 15,000 plants) of showy lady's slipper orchids. It is located northwest of McDonalds Corners. The orchids typically flower in June, but the exact time depends upon the weather. There is a boardwalk with a brochure to guide you. Please remember to stay on the boardwalk to protect the rare plants.

If you visit, watch for other unusual plants such as pitcher plants and sundews. Since they grow in peat, most plants in fens find it very difficult to gather enough nutrients to grow. Some fen plants are evergreen, and this allows them to conserve nutrients by using the same leaves for many years. pitcher plants and sundews, however, solve this problem a different way. They trap insects, obtaining nitrogen and phosphorus from their corpses rather than from the soil.

The Purdon property was bought and protected for your use by a unique conservation group — the Nature Conservancy of Canada. The Nature Conservancy exists for only one purpose — to buy and protect important natural areas. Here is a conservation group that puts their money where their mouth is. They depend entirely upon donations from the public and private industry to buy areas like the Purdon fen. You may wish to consider adding them to the charities you support. Their address is given near the end of this book.

The K&P Trail (Area 7)

The Kingston and Pembroke railway was chartered in 1871.[40] Owing to the difficulty of the terrain, it never extended past Renfrew. It took 12 years to build only 112 miles. Contrast this with the CPR which was extended from the Great Lakes to the Pacific Ocean in only ten years (1875-1885)! A principal function of the K & P railway was to carry pine and other lumber southward to Kingston for export to the United States. Lavant, Folger, Clyde Forks and Flower Station were some of the settlements along the K & P. The principal environmental impact of the line was the increased destruction of forests through logging, fire and human settlement. The K & P Railway was eventually abandoned as a railway and it is now a recreation trail. It is a good way to see some of the rugged northern lands of Lanark County. If you visit Kingston, you can still see the railway station that was the southern terminus of this line. It is presently a tourist information office.

Lavant/Darling Spillway (Area 9)[41] (Area 8 covered under Area 3)

The remains of an ancient river (a spillway), snake through Lavant and Darling townships (Map 16). This river may have flowed northeast about 10,000 BP through what is now White Lake into the Champlain Sea. White Lake could be considered just a small puddle from this glacial river. Most of this old river bed is now dry, or has only small streams or wetlands in the bottom. Geographers call such small streams misfit streams because they are far too small to have carved out the large valleys they now occupy. Some of our finest forests perch in deep ravines along the edge of this spillway and with luck you may see Red-shouldered Hawks flying overhead.

At one point Highway 511 plunges down into the spillway. A better tour comes on a side road going west from Highway 511 to Clyde Forks; watch for it about 6 kilometres south of the Lanark-Renfrew Boundary. This road follows one branch of the old spillway and wanders up and down its side showing both valley bottom and some fine views from hill sides. Some of these areas were once cleared for agriculture but most are modestly forested. Some of the forest was being logged as the first edition of this book was written; a few fine pieces of old forest remain. One of the best is in deep ravines on the east side of this road about a kilometre west of the Lavant-Darling boundary. It is not a good area to explore on foot because the soil on the steep slopes is easily eroded, but it offers a spectacular view from the car.

As you drive through this landscape it may be difficult to imagine the volumes of water that would have carved it, or to picture them being produced by the melting ice sheet. Remember that the spring flooding we see today comes from a few metres of snow melting. After the ice age there were kilometres of ice melting. So picture a rapid spring thaw, and try multiplying it by a thousand or so, and then extend it over a few centuries!

White Lake (Area 10)

White Lake is in the northern part of Lanark County. Surrounded by Precambrian rocks, the lake is more like those one finds in Algonquin Provincial Park. The lake is not entirely natural because water levels have been altered by a dam on the east end. This has modified many shoreline areas and produced beds of drowned trees. Peat bogs and cat-tail marshes can be found in some of the bays. Wetlands on the east side of the lake, designated provincially significant and an Area of Natural and Scientific Interest, include large areas of fen, a relatively uncommon kind of wetland.

The lake is best explored by boat. A public launch area provides access to the west end of the lake from Highway 511. The road to this launch area passes through some white pine forests with large boulders. Some boulders near the boat launch approach the size of small cars and are good reminders of the power of moving ice. Boats can be rented at the east end of the lake in the town of White Lake.

A small fire barren occurs on the north shore of the lake. It has typical fire resistant trees such as red oak, as well as unusual plant species such as wild sunflower.

Pakenham Mountain (Area 11)

Although perhaps not as dramatic as Wolf Grove (Area 15), Pakenham Mountain is another outcropping of Canadian Shield that would also have been a rocky promontory in the Champlain Sea. Like Wolf Grove, this high and rocky country was bypassed by early settlement roads, and so it still retains its forests. One cannot drive through it, but one can drive around it. An interesting trip can be taken by starting in Pakenham and driving clockwise around Pakenham Mountain, passing through Cedar Hill and by some of shores of White Lake along the way. This keeps the high shield outcropping on one's right. A little more than ten kilometres out of Pakenham (after Cedar Hill), one can actually see a bluff cut into the hill side, apparently by the old Champlain Sea. Here one is about 500 feet above present day sea levels. These sandy eroded areas still support distinctive plants which tolerate dry south-facing slopes. The complex of wetlands on Pakenham Mountain is designated provincially significant, but the area deserves further conservation protection status. Perhaps some day there will be a walking trail through it, and a viewing platform for looking eastward across the Ottawa Valley from its summit.

Pakenham Bridge (Area 12)

The Pakenham Bridge is of historic interest, but it also provides a chance to see the Mississippi River as it rushes over exposed limestone bedrock on its way out of Lanark County. The Mississippi River starts in Frontenac County and flows northeast through Lanark County. It begins in the rugged shield country with extensive forests, deep lakes and boulder-strewn streams. I put a hole in my canoe many years ago in one such stream. But by the time the Mississippi reaches Pakenham, it is flowing through agricultural land, deep clay soils and flat limestone outcrops. The river has an entirely different personality here. Not only is it more placid, but it is carrying clay from eroding agricultural land and nutrients from sewage treatment plants in towns upstream like Carleton Place and Almonte. The view here is pleasant, and there is a picnic area. But the slippery green growths of algae on the rocks remind us that Lanark County is not immune to pollution. Cottages, towns and farms far upstream are changing the water quality of a river that the early settlers could drink from. This is not recommended now. The rock outcrops at Pakenham bridge are an Area of Natural and Scientific Interest.

Mill of Kintail (Area 13)

Many small streams rise in the western shield country and then drain east through clay plains and limestone. The Mill of Kintail Conservation Area provides access to one of these streams. The mill itself is a reminder of how much human history is connected to the water power provided by streams. The stream valley itself is cut down to limestone bedrock, and is a good place for seeing stream bank plants. Unfortunately, it is also a good place for seeing large stands of purple loosestrife, the European wetland plant which has now invaded many of our wetlands. It is probably displacing many of our native streamside plants. Seeing the explosion of purple loosestrife provides a blunt reminder of how dangerous it is to introduce plants and animals that are not native to this part of the world. Across the stream from the conservation area are some fine forests to explore.

Burnt Lands Alvar (Area 14)[42]

In a county of forests and wetlands there is at least one area which is like prairie. The Burnt Lands Alvar is on County Road 49 east of Almonte on the east side of Ramsay Township. An alvar is an unusual plant community that develops where there is limestone bedrock with such shallow soil that forests cannot grow. Flooded in the spring, droughted in mid-summer, this is a harsh place for anything to grow. As a consequence many unusual plant species occur here, including a provincially rare panic grass. Flowering asters, bluebells, and goldenrods carpet many areas in mid-summer. The unique plants mean that other unusual species are found here including rare butterflies. Scientists have studied this alvar and compared it to other Ontario alvars; they have concluded that it is one of the best examples left in the province. Other alvars are found in the eastern United States and in Scandinavia, but they have different kinds of plants in them. The Burnt Lands Alvar therefore is a very special part of Lanark County.

The number of unusual plant species, and the fact that so many of them need open sunny conditions, tells us that part of this area was probably never forested. In some areas there is no soil — just bare exposed limestone and moss. Presumably such areas were always open and sunny, and the rare plants may have survived on the edges of such clearings. Fire may have helped maintain the clearings, and there seems to be little doubt that large fires in the last century have increased the size of the alvar. Of course, we must not forget that natural fires occurred in the forests of this area and played a role as well. The name of this site apparently originates in the great fire of 1870 (Map 10).

Owing to its importance, the Ontario Ministry of Natural Resources has designated 1,500 ha of this alvar as an Area of Natural and Scientific Interest; within it, 127 ha comprise the Burnt Lands Provincial Park (Nature Reserve). In this they have the full support of professional scientists who have studied the area. ANSI designation gives the land little legal protection, but it does encourage both the actual landowners and the Township Council to take responsibility to protect its natural and scientific values. A significant portion of this unique area is on private land, although several hundred acres have been owned for many years by the Ontario Ministry of Natural Resources. Unfortunately, it is difficult to appreciate that this is a provincially significant alvar for two reasons. First, county residents have used it as a garbage dump, so that rare plants often emerge from beneath rotting mattresses and old lumber. Second, the OMNR planted some of the area with jack pine trees some years ago. These trees are now shading out the rare plants. Perhaps a group of Scouts or other public service group will volunteer to cut them all down before the rare plants disappear. A large quarry also threatens another part of the alvar. Since much of the ANSI is private land and trespassers are unwelcome, be certain to check your maps carefully. We can hope that the OMNR will eventually arrange a parking area, paths and a display where you can visit the site and learn more about alvars.

Wolf Grove (Area 15)

A great peninsula of Precambrian rock protrudes eastward out of the Canadian Shield, pointing like a finger at Almonte. This outcropping of hard rock proved to be such an obstacle to road construction that few concession roads cross it. It was also little developed for agriculture, so it remains as a fortress of forests and beaver ponds surrounded by agricultural land. Although early roads bypassed this area, one can now drive almost straight through Wolf Grove by taking Highway 16 west from Almonte toward Middleville and Hopetown (Highway 16 is a paved westward extension of County Road 49 through Almonte). A little more than a kilometre out of Almonte one climbs out of good agricultural land on clay plains into rocky, forested Canadian shield country. You are in Wolf Grove. If you turn left (south) about ten kilometres out of Almonte, and then take your next left (after about two kilometres), you can return back toward Almonte. Here you will find yourself on the old Perth Road, which winds from Almonte to Perth. This road will take you along the southern limit of Wolf Grove. Notice the rocky hills and forest to the north, and the better agricultural land to the south. Much of this road also follows the old shoreline of the Champlain Sea. Imagine as you drive along, seals basking on the rocks and sea birds flying overhead. At present, 255 ha of wetlands in Wolf Grove are designated provincially significant (Map 14), but more of this island of Canadian Shield and former peninsula in the Champlain Sea deserves conservation status and protection.

Old Perth Road (Area 16)

There are no markers for this sea shoreline, and it takes a practised eye to pick it out. However, early settlers who were constructing roads had an eye for flat sandy soils, which often happen to be old beaches. The old Perth Road is therefore a good route to follow. Drive the section south from Almonte, and it will bring you out, after a winding 15 kilometres or so, to Boyds Settlement; here you can jog to Highway 7 for your return trip. If you follow this route, you will meander along land which is about 500 feet above sea level, the approximate contour line of the old sea. On your right you will first see the rocky outcrop that marks the southern boundary of Wolf Grove, and then you will come to sandier areas. If you keep a sharp eye out, you will occasionally see old beach lines, which look like steep shelves cut into the hills on your right. If you turn back east along Highway 7 toward Carleton Place, you will drop down over the old shoreline of the Champlain Sea again as you pass Ramsay Concession Road 2.

Appleton Silver Maple Swamp (Area 17)

For thousands of years the Mississippi River has flooded each spring as water poured in from melting snow and spring rain. Swamps and marshes all along the river depend upon this annual flood. Between Appleton and Almonte there is an extensive silver maple swamp that has formed where silt and clay have settled out from the annual floods. silver maple is one of our most flood-tolerant tree species. It can survive without flooding — we know this because silver maple trees are often planted in parks. But in natural situations, silver maple grows only where it is too wet for other trees to survive (recall Figure 1). The combination of flooding and shade from the trees prevents all but a few hardy herbaceous plants from growing here. One is the ever-present sensitive fern, another is poison ivy. In the shallow water there are fine beds of aquatic plants. This provincially significant wetland and Area of Natural and Scientific Interest is an excellent place to visit, but is only accessible by canoe. Remember that the surrounding land is private.

Although this may seem remote and undisturbed, areas like this throughout eastern Ontario are threatened precisely because they depend upon spring floods. People appear unwilling to accept the reality that melting snow produces spring floods, and therefore tend to build cottages and homes where the river would naturally flood each year. It seems that by August, humans have already forgotten the water levels from the preceding spring. This means that dams have been built to constrain spring flooding on the Mississippi River. As floods are controlled, the area of swamps and marshes slowly shrinks. We can hope that current restrictions which prevent further building on floodplains will be enforced strictly so that spring floods and silver maple swamps will be here for generations to come.

The Edge of the Shield (Area 18)

Just west of Carleton Place you can pass a boundary that marks more than a billion years of the earth's history, as well as being one of the most important ecological boundaries in eastern Ontario (see Map 2). Drive west on Highway 7 from the lights at the Carleton Place junction (Highway 15). Note that the land is fairly flat. Watch for rock cuts on the edge of the road, and note that the rocks are limestone — a greyish-white rock with discrete layers. This part of Highway 7 is still on the Smiths Falls limestone plain. (For an even better feel for this limestone landscape, drive south from the Carleton Place junction on Highway 15.) After driving west of Carleton Place on Highway 7 for about 5 kilometres, the road starts to climb, and the rocks become jagged and pinkish-grey. You have now climbed onto the Canadian Shield. Road work in 1999 exposed fresh new rocks, making the difference even more notable.

If you continue a few kilometres farther west on Highway 7, just past Ramsay Concession Road 2, the rock outcrops on the north side of the road become crisp white, without the horizontal layers found in limestone. Here is the first exposure of marble. Once you recognise it, you can follow the exposures of marble in rock cuts along side roads by watching for the white crystalline rocks. The best exposed marble is farther north in Darling Township. A large marble quarry near Tatlock has conveniently placed huge chunks of marble on the edge of the road. The roadside is a good place for children to get marble chips for rock collections. Some of the nearby road surfaces are white because gravel from the quarry is used on them.

The Carleton Place Hackberry Stand (Area 19)

Perched on the limestone beside the rapids at Carleton Place is a population of rare southern trees — hackberry. These trees have distinctive bark that is smooth with strong ridges. No one really knows how these trees spread here — does it perhaps go back to a time when Carleton Place was on the shoreline of the Champlain Sea? There is also evidence that there was a brief warm period lasting from about 8,000 to 4,000 years ago. This was the time when many other hardwood trees moved back from the south to Lanark County. The climate has been one to two degrees cooler since then. Perhaps hackberry trees moved into Lanark County during this warm period, and since have survived only in Carleton Place. But why they should survive there is difficult to explain. Perhaps the mist from the rapids helps prevent occasionally short killing frosts; in much the same way commercial orange plantations in Florida will spray water on the trees during cold spells.

Mississippi Lake (Area 20)

We may think of Mississippi Lake as being a small fresh water puddle left over from the Champlain Sea. It is noteworthy because it has formed right where the Canadian Shield meets the Smiths Falls limestone plain. The west shore is Precambrian shield, the east is part of the Smiths Falls limestone plain. It therefore has a sort of split personality — some sections remind one of the limestone lakes farther south, and others seem more typical of the shield lakes of the north. There are excellent beds of aquatic plants to explore, as well as some fine areas of silver maple swamp in bays. If you rent a canoe, you can explore these areas easily. With care, you can see both Great Blue Herons and Ospreys fishing. These species nest many kilometres west of the lake in isolated beaver ponds where they are not disturbed by people. Loons may be seen too, but unlike Great Blue Herons and Ospreys, they nest right on the shoreline, and are increasingly vulnerable to disturbance from motor boats and cottages. During June, one can see several species of turtles including painted turtles and snapping turtles emerging to sun themselves on logs or nest on beaches. With luck you may see a rarer species such as Blanding's turtles or a musk turtles. Unfortunately, all these species are being harmed by the development of the beaches and shorelines where they lay their eggs. During migration the lake is a fine place for observing waterfowl including large flocks of Canada Geese.

Each April the pickerel of Mississippi Lake are involved in a mass migration that can be observed in Innisville where the Mississippi River flows into Mississippi Lake. Highway 7 conveniently passes right through the middle of Innisville. Pickerel, like many other fish, must lay their eggs in gravel covered with swiftly flowing water. A single female may lay up to a million small eggs. Once fertilised by males, these eggs fall into the crevices of the gravel for protection, and hatch about two weeks later. Entire populations of fish may therefore be dependent upon such small spawning areas. These pickerel are currently in decline because of over-fishing and the destruction of fish habitat in Mississippi Lake, but the Innisville pickerel run is still worth a spring visit. The wetlands at Innisville have been designated provincially significant and are incorporated in an Area of Natural and Scientific Interest.

Moose were once common in Lanark County, but declined as the forests were cleared, and as deer invaded from further south. Deer carry a parasite that is fatal to moose, and so long as deer populations remain high, moose will likely remain scarce.

The Northern Forests

Several roads snake through the northern forests of Lavant, Darling and Dalhousie townships. To see them, try driving north on Highway 511 from Lanark, or west on County Road 16 through Poland and Lavant. A substantial part of this is Crown Land. That is, it is owned by the public and managed for us by the Ontario Ministry of Natural Resources. This is one of the few areas of eastern Ontario where large portions of the land remain in public ownership. The future of these forests is currently being debated. Almost all of it has been logged over in the past, and is scheduled to be logged over again by private logging companies. Some groups, such as the Canadian Parks and Wilderness Society, want to see some examples of these forests set aside to recover from past logging. Currently, the Darling Township ANSI is the only protected area. Given sufficient time, the forests in such protected areas will regrow into the splendid forests that once covered most of Lanark County. Many citizens agree that more forest lands should be protected, particularly those that are most similar to our old forests, or those with rare plant and bird species. Other groups strongly oppose this view, largely out of fear that this will interfere with traditional uses such as hunting and logging. You may consider your views on the future of these forests as you explore them. If you are interested, the address for the Canadian Parks and Wilderness Society is also given near the end of this book.

Shrike Nesting Area

Loggerhead Shrikes are a bird species now on the endangered species list in Ontario. They breed in open prairie-like conditions such as those found in the limestone plains of Montague Township east of Smiths Falls. These prairie-like conditions are disappearing in Lanark County as forests grow back and as estate lots replace fields. We do not show exactly where these birds breed because it is important that they not be disturbed. If you see any on your travels, be sure to notify the Ontario Ministry of Natural Resources.

The Holleford Meteor Crater

Just southwest of Lanark County lies an ancient meteor crater more than 2 km in diameter. Humans are only beginning to appreciate the role of collisions with meteors in the history of life on Earth. It now seems increasingly clear that mass extinctions such as the disappearance of the dinosaurs can be traced back to such cataclysmic events. It is one thing to read about collisions, or watch a Hollywood movie about them; it is quite another to stand on the edge of a real crater that was gouged and melted into Precambrian rocks. This crater can be reached by driving southwest either to Verona or Sydenham; it lies almost exactly midway between these settlements, just inland from Knowlton Lake on the Portland-Loughborough township line. A side road through Holleford provides a dramatic view of the crater and a blue historical plaque helps identify the site. The crater is sufficiently old that it has been partially filled with Palaeozoic sediments dating from some quarter of a billion years ago. Fossils of Palaeozoic invertebrates litter the edge of the road and illustrate that at one time this crater lay under an ocean teeming with marine life.

Enthusiasts may then wish to drive north long the Portland-Loughborough township line. About 1 km north of Holleford one crosses a road cut where you can plainly see rock layers that have been violently twisted and warped. Originally horizontal layers of rock appear to even have

The Holleford meteor crater may have looked something like this about 250 million years ago, although at other times it was probably submerged under a shallow ocean.

been flipped vertically onto edge. These deformed rocks illustrate the violence of meteoric impacts. Perhaps some amateur geologist will prepare a guide book to the entire area, but even lacking this, a family excursion to the Holleford Crater is bound to be memorable.

Fishers, a large member of the weasel family, were once nearly wiped out by the destruction of forests and over-trapping. Now they are returning to the county, a positive sign of improving environmental conditions. They are one of the principal predators on porcupines.

Conclusion

The landscape of Lanark County has changed dramatically in a little over ten thousand years. Solid ice, barren gravel and the Champlain Sea gave way in turn to magnificent hardwood forests. Then, in a little over two centuries these forests were cleared by settlers and farms and cities established. One of Canada's largest urban areas is rapidly expanding just to the east, yet Lanark County still retains much of its natural beauty.

I hope that this slim volume has allowed both visitors to the county and life-long residents to more fully appreciate the natural environment here. Forested lands are precious, and even if we live in cities, we need the air, water, food, fuel and wildlife that they generate.

Where does the future lie? How many people should live in Lanark County? How big should our towns get? How much land will we protect for agriculture and forestry? Which areas will we set aside as protected wetlands and forests? Compared to many parts of the world, we have few natural areas with legal protection. If tiny and relatively poor countries like Costa Rica can set aside forest reserves and parks, surely Lanark County can too. We could imagine a future where the best agricultural land is set aside to ensure food supplies for our children, and where urban sprawl is limited. We could imagine some large protected wetlands, and a large protected area of forest which is being allowed to gradually grow back to remind us of the kinds of forest which once covered the county. We will not see them in our lifetimes, but our grandchildren might — and thank us for our foresight. We could imagine a thriving tourist economy attracting citizens from the Regional Municipality of Ottawa-Carleton. Other areas could be managed to yield high grade forestry products to keep our mills running and our economy thriving. Perhaps the forest economy could expand to a more complete mixture of maple syrup, furniture, medicinal plants, fuel wood and quality lumber, all processed in our own communities. Healthy forests can mean healthy logging industries and busy tourism, as long as we leave room for both. Achieving a wise mixture of uses requires that we first understand where we are and what we have to work with. Thus this book. What remains is for us to exercise some forethought, and to work together as a community to make the most of our natural environment in Lanark County.

Further Reading and References

Anderson, T.W. 1989. Vegetation changes over 12 000 years. Geos (3) 39-47.

Bennett, C. and D.W. McCuaig. 1981. In Search of the K & P. Juniper Books Ltd., R.R. 2, Renfrew.

Bennett, C. and D.W. McCuaig. 1982. In Search of Lanark. Juniper Books Ltd., R.R. 2, Renfrew.

Beschel, RE., PJ. Webber and R Trippett. 1962. Woodland transects of the Frontenac Axis region, Ontario. Ecology 43: 386-396.

Bowering, I. (ed.). 1992. Bowering's Guide to Eastern Ontario. Quarry Press, Kingston.

Cadman, M.D., D.A. Sutherland, G.G. Beck, D. Lepage and A.R. Couturier (eds.). 2007. Atlas of the Breeding Birds of Ontario, 2001-2005. Bird Studies Canada, Environment Canada, Ontario Field Ornithologists, Ontario Ministry of Natural Resources and Ontario Nature, Toronto.

Chapman, L.J. 1975. The Physiography of the Georgian Bay-Ottawa Valley Area of Southern Ontario. Ontario Division of Mines, Geoscience Report 128. Accompanied by Map 2228, scale 1 inch to 4 miles or 1:253,440.

Chapman, L.J. and D.F. Putnam. 1984. The Physiography of Southern Ontario. Third edition. Ontario Geological Survey, Special Volume No.2. Government of Ontario, Toronto.

Davis, M.B. 1976. Pleistocene biogeography of temperate deciduous forests. Geoscience and Man, Volume XIII (March 15) 13-26.

Eyles, N. 2002. Ontario Rocks: Three Billion Years of Environmental Change. Fitzhenry & Whiteside, Markham.

Gentilcore, R.L. and C.G. Head. 1983. Ontario's History in Maps. University of Toronto Press, Toronto.

Hall, R.I. 1993. Paleolimnological analysis of lake-watershed interactions and long term lake trophic status. Ph.D. Thesis, Department of Biology, Queen's University, Kingston.

Head, C.G. 1975. An introduction to forest exploitation in nineteenth century Ontario. Pages 78–112 *in* J.D. Wood (ed.), Perspectives on Landscape and Settlement in Nineteenth Century Ontario. The Carleton Library No. 91, McClelland and Stewart Limited, Toronto in association with the Institute for Canadian Studies, Carleton University.

Howe, C.D. 1915. The effect of repeated forest fires upon the reproduction of commercial species in Peterborough County, Ontario. p. 116-211 *in* Forest Protection in Canada, 1913-1914, Commission of Conservation of Canada, William Briggs, Toronto.

Hughson, J.W. and C.C. J. Bond. 1965. Hurling Down the Pine. The Historical Society of the Gatineau, Old Chelsea, Quebec. First edition 1964, Revised second edition 1965.

Keddy, P.A. and C.G. Drummond. 1996. Ecological properties for the evaluation, management, and restoration of temperate deciduous forest ecosystems. Ecological Applications 6: 748-762.

Keddy, C.J. 1993. Forest History of Eastern Ontario. Prepared for the Eastern Ontario Model Forest Group, Kemptville.

Lockwood, GJ. 1991. Beckwith: Irish and Scottish Identities in a Canadian Community. Beckwith Township, R.R. 2, Carleton Place.

Peterson, R.T. 2002. A Field Guide to the Birds of Eastern and Central North America. Houghton Mifflin, Boston.

Weller, W.F. and MJ. Oldham (eds.). 1988. Ontario Herpetofaunal Summary 1986. The Ontario Conservation Data Center, Peterborough.

White, D. 2008. Plants of Lanark County, Ontario (revised edition). www.lanarkflora.com (accessed 19 August 2008).

Some Other Sources of Information and Guided Tours *(Updated 2021)*

Canadian Parks and Wilderness Society, Ottawa Valley Chapter
15 Rue Taschereau #240
Gatineau, Québec J8Y 2V6
phone: 819-778-3355
www.cpaws-ov-vo.org

Friends of Murphys Point Park
2243 Elm Grove Rd.
Perth, Ontario K7H 3C7
www.friendsofmurphyspoint.ca

Kingston Field Naturalists
P.O. Box 831
Kingston, Ontario K7L 4X6
www.kingstonfieldnaturalists.org

Lanark County Tourism Association
99 Christie Lake Rd.
Perth, Ontario K7H 3C6
phone: 613-267-4200 ex. 1530
www.lanarkcountytourism.com

Macnamara Field Naturalists' Club
P.O. Box 391
Arnprior, Ontario K7S 3L9
www.mfnc.ca

Mississippi Madawaska Land Trust
10970 Highway 7
Carleton Place, Ontario K7C 3P1
phone: 613-253-2722
www.mmlt.ca

Mississippi Valley Conservation Authority
10970 Highway 7
Carleton Place, Ontario K7C 3P1
phone: 613-253-0006
www.mvc.on.ca

Mississippi Valley Field Naturalists
P.O. Box 1617
Almonte, Ontario K0A 1A0
www.mvfn.ca

Murphys Point Provincial Park
2243 Elmgrove Road
Perth, Ontario K7H 3C7
phone: 613-267-5060
www.ontarioparks.com/park/murphyspoint

Nature Canada
300-240 Bank St.
Ottawa, Ontario K2P 1X4
phone: 613-562-3447
www.naturecanada.ca

Nature Conservancy of Canada
350 Sparks St.
Ottawa, Ontario K1R 0A4
phone: 800-465-0029
www.natureconservancy.ca

Ontario Ministry of Natural Resources and Forestry
10 Campus Dr.
Kemptville, Ontario K0G 1J0
phone: 613-258-8204
www.infogo.gov.on.ca/infogo/#orgProfile/1395/en

Ottawa Field-Naturalists' Club
P.O. Box 35069, Westgate Post Office
Ottawa, Ontario K1Z 1A2
www.ofnc.ca

Endnotes

1. Shilts, W.W. and I.M. Kettles. 1989. Geology and acid rain in eastern Canada. Geos 18(3):25 32.

2. Anderson, T.W. 1989. Vegetation changes over 12 000 years. Geos (3): 39-47.

3. Chapman, L.J. and D.F. Putnam. 1984. The Physiography of Southern Ontario. Third Edition. Ontario Geological Survey, Special Volume No.2. Government of Ontario, Toronto.

4. Barnes, B.V. 1991. Deciduous forests of North America. *in* E. Rohig and B. Urlich (eds.), Ecosystems of the World 7: Temperate Deciduous Forest. Elsevier, New York, NY.

5. Braun, E.L. 1950. Deciduous Forests of Eastern North America. The Blakiston Co., Philadelphia, PA.

6. Lockwood, G.J. 1991. Beckwith: Irish and Scottish Identities in a Canadian Community, 1816-1991. Beckwith Township, R.R 2, Carleton Place.

7. Ibid.

8. Brown, H.M. 1984. Lanark Legacy: Nineteenth Century Glimpses of an Ontario County. Corporation of the County of Lanark.

9. Keddy, CJ. 1993. Forest History of Eastern Ontario. Eastern Ontario Forest Group, Kemptville.

10. Lockwood, G.J. 1991. Ibid.

11. Anderson, T.W. 1985. Late-quaternary pollen records for eastern Ontario, Quebec, and Atlantic Canada. p. 282-326 *in* V.M. Bryant, Jr. and RG. Holloway (eds.), Pollen records of late-quaternary North American sediments. AASP Foundation, Dallas, TX.

12. Davis, M.B.. 1976. Pleistocene biogeography of temperate deciduous forests. Geoscience and Man XIII (March 15): 13-26.

13. Keddy, P.A. and P. MacLellan. 1990. Centrifugal organization in forests. Oikos 59: 75-84.

14. Martin, P.S. and H.E. Wright. 1967. Pleistocene Extinctions: The Search for a Cause. Yale University Press, New Haven, CT.

15. Hibbard, C.W., D.E. Day, D.E. Savage, D.W. Taylor and J.E. Guilday. 1965. Quaternary mammals of North America. p. 509-525 *in* H.E. Wright, Jr. and D.G. Frey (eds.), The quaternary of the United States. Princeton University Press, Princeton, NJ.

16. Day, G.M. 1953. The Indian as an ecological factor in the northeastern forest. Ecology 34:329-346.

17. Day, G.M. and B.G. Trigger. 1978. Algonquin. p. 792-797 *in* B.G. Trigger (ed.), Handbook of North American Indians: Volume 15 Northeast. Smithsonian Institution, Washington, DC.

18. Jamieson, J.B. 1990. The archaeology of the St. Lawrence Iroquians. p. 385-404 *in* C.J. Ellis and N. Ferris (eds.), The archaeology of southern Ontario to A.D. 1650. Occasional paper of the London Chapter, OAS Number 5.

19. Lower, A.R.M. 1929. The assault on the Laurentian barrier, 1850-1870. Canadian Historical Review 10:304.

20. Parson, H.E. 1987. The colonization of the southern Canadian Shield in Ontario: the Hastings road. Ontario History LXXIX:263-273.

21. Brown, H.M. 1984. Ibid.

22. Lockwood, G.J. 1991. Ibid.

23. McGill, J.S. 1968. A Pioneer History of the County of Lanark. T.H. Best Printing Company, Toronto.

24. Gentilcore, RL. and C.G. Head. 1983. Ontario's History in Maps. University of Toronto Press, Toronto.

25. Ibid.

26. Bennet, C. and D.W. McCuaig. 1981. In Search of the K & P. Juniper Books Ltd., R.R. 2, Renfrew.

27. Howe, C.D. 1915. The effect of repeated forest fires upon the reproduction of commercial species in Peterborough County, Ontario. p.116-211 *in* Forest Protection in Canada, 1913-1914. Commission of Conservation of Canada. William Briggs, Toronto.

28. Ibid.

29. Walker, H.J. 1967. 100 Years Ottawa and the Valley. A Backward Glance from Centennial Year. Reprinted from the Ottawa Journal, January-March, 1967. p.48.

30. Billings, C.E. 1910. The Great Fires of 1870. Women's Canadian Historical Society of Ottawa 3:27-35. p.29.

31. Ibid. p. 27.

32. Ibid. p. 29.

33. Walker, H.J. 1967. Ibid. p. 48.

34. Cadman, M.D., P.F.J. Eagles and F.M. Helleiner. 1987. Atlas of the Breeding Birds of Ontario. University of Waterloo Press, Waterloo.

35. Cadman, M.D., D.A. Sutherland, G.G. Beck, D. Lepage and A.R. Couturier (eds.). 2007. Atlas of the Breeding Birds of Ontario, 2001-2005. Bird Studies Canada, Environment Canada, Ontario Field Ornithologists, Ontario Ministry of Natural Resources and Ontario Nature, Toronto.

36. Ibid.

37. White, D. 2008. Plants of Lanark County, Ontario (revised edition). www.lanarkflora.com (accessed 19 August 2008).

38. Hall, RI. 1993. Paleolimnological analysis of lake-watershed interactions and long term lake trophic status. Ph.D. Thesis, Department of Biology, Queen's University, Kingston.

39. Ibid.

40. Bennet, C. and D.W. McCuaig. 1981. Ibid.

41. Chapman, L.J. and D.F. Putnam. 1984. Ibid.

42. Belcher, J., P.A. Keddy and P.M.C. Catling. 1992. Alvar vegetation in Canada: a multivariate description at two scales. Canadian Journal of Botany 70: 1279-1291.

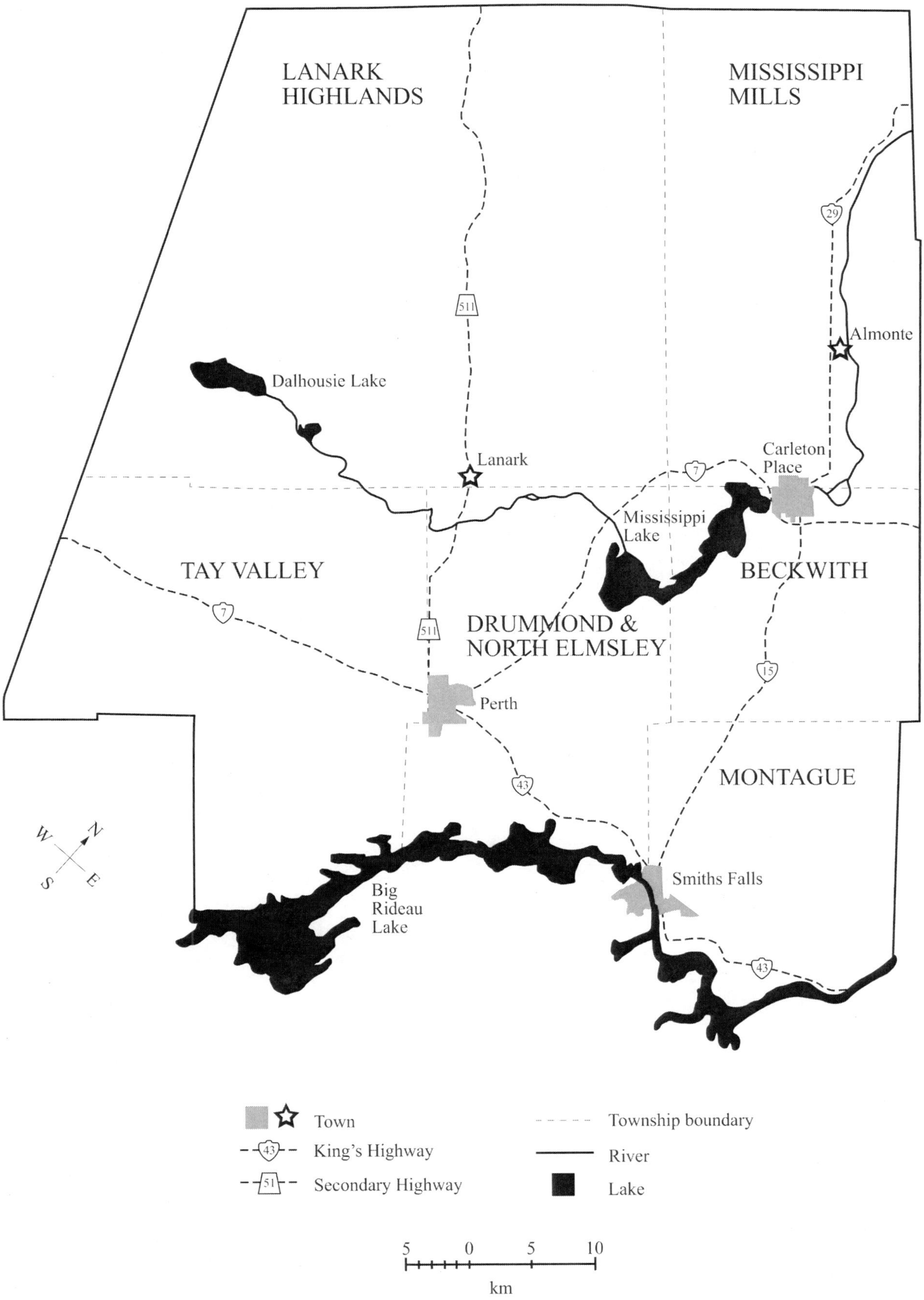

Map 19. Major municipalities of Lanark County as of 2008.

APPENDICES

Appendix A

Green Gems: A Conservation Priority List for Lanark County

One of the most beautiful aspects of life in Lanark County is our natural environment. Much of southern Ontario has already been reduced to subdivisions and cities, surrounded mostly by fields of corn. Many people now live crowded into large cities. There are few places in the world where one can live surrounded by wild landscapes. Here in Lanark County, we still have that good fortune. It will take some common sense to keep it that way.

One common sense action is to ensure that the most important wild places in our county are recognized and protected. These places are Lanark County's green gems.

There is an international initiative to protect important wild places around the world, the United Nations' Strategic Plan for Biodiversity. Big wild places like Algonquin Provincial Park and the Great Barrier Reef are well-known examples, but they are far away. They can also be expensive to visit. What about smaller places that are closer to home?

One familiar green gem closer to home is Murphys Point Provincial Park. It has some fine wetlands and forests and an important population of the threatened Gray Ratsnake.

Other local green gems are not so well-known. Here is one example. Every time you cross the Highway 7 bridge at Innisville, you are passing through an enormous wetland, including a vast silver maple swamp. The view from Highway 7 is mostly obstructed by buildings. Yet, downstream from the bridge is the Mississippi Lake National Wildlife Area, 264 hectares of mostly wetland, and the McEwan Bay Provincially Significant Wetland. The open marsh and shallow water provide habitat for migrating waterfowl. Upstream from the bridge is a larger area of Mississippi River floodplain, the Steward Lake-Haley Lake Provincially Significant Wetland. Together these areas form the Innisville Wetlands Area of Natural and Scientific Interest (ANSI), an official designation for a wild place that has been carefully studied by professional biologists and found to be of provincial value.

Although the Innisville Wetlands ANSI is surely a green gem, it seems to have a low profile in the county. Most people I meet are unaware of its existence or importance.

Lanark County has many other important natural areas that remain virtually unknown in our community. What might some of them be?

When I wrote my guide to the natural environment of Lanark County back in 1999, it included a short list of some of our important natural areas—our "Special Places." The list included Murphys Point Provincial Park, Purdon Conservation Area, and Burnt Lands Alvar. Twenty years have passed. I have done more talking to local naturalists and biologists and new information has become available.

One example of new information is our county flora that has been compiled by David

White. His web site *Plants of Lanark County, Ontario* (www.lanarkflora.com) names all our plant species and describes where they are found. This site also describes 17 important areas that have exceptional plant diversity. The Burnt Lands Alvar, for example, has plants found nowhere else in Lanark County.

We also have seen new efforts to identify and protect wild places. The Mississippi Madawaska Land Trust now exists to accept donations of ecologically significant land for protection in perpetuity. MMLT protects wild places including Blueberry Mountain and Poole Family Nature Sanctuary.

At the same time, development continues to sprawl westward from Ottawa. Bulldozers continue to destroy our forests and wetlands. Meanwhile, our county and townships are slowly moving toward plans to guide future land use in Lanark County. One planning responsibility is to ensure that our vital areas of wildlife habitat remain.

So, this is a good time to ask: where are these vital areas? How many are there? What kinds of unusual plants and animals occur in them? We need to know. Such a list of our vital green spaces would provide us with a precious kind of information: the areas of our county that we must protect for future generations.

Since the county does not have a plan to describe or protect such areas, I am going to share my personal list and a map. I started with the list of our special places in my own original guide, *Earth, Water Fire: An Ecological Profile of Lanark County*. I then added information from David White's county flora. Further, I incorporated some sites, such as Provincially Significant Wetlands and Areas of Natural and Scientific Interest, for which I had other information collected by professional biologists. This is not a full list, but it shows some top priorities—our green gems.

A few are obvious. Murphys Point Provincial Park, the Purdon Conservation Area and Blueberry Mountain, for example. These are all familiar gems, already known and appreciated. All three have trails for visitors. More, you ask? The Innisville Wetlands, mentioned above, is on the list. And, of course, so is the Burnt Lands Alvar.

Before I say more about the other locations, let me offer a few cautions.

There are risks in naming the county's special places. One risk is that it will attract the wrong kind of people, such as those who dig up wild orchids. It would be wrong to let such people know where certain rare orchids occur. Other people might want to see certain rare birds, but by visiting the area in May or June, they might interfere with nesting. Other people want to collect wild plants to eat, even if those plants should not be harvested. So, caution is necessary.

Another reason for caution is that some green gems are privately owned. Green gems need not be government land and they need not be open to the public for recreation. Sometimes wild nature needs to be left alone. Our own property, Keddy Nature Sanctuary, for example, is protected by the Mississippi Madawaska Land Trust through a conservation agreement. This agreement will protect the forests and wetlands in perpetuity. Visitors can tour the property by appointment only, or during guided walks. One reason for this limitation is to protect wild birds while they breed.

So, there is potential harm in naming our important wild places. But there is also a problem with silence. If we don't name wild places, they can be forgotten or ignored. If our planners don't put them on maps then the next thing you know they are logged or subdivided and gone forever. In 2017 our own county chose to side-step environmental protection measures for numerous candidate ANSIs and handed this responsibility off to individual townships. How are our townships supposed to plan for the future and protect important natural habitats, if they don't even have documentation of where these important areas are? Townships don't have the budget to hire teams of professional biologists. So, for this reason, someone needs to speak on behalf of our green gems.

Why me? Mostly because if I don't write it, I doubt anyone else will. I am a professional biologist, I went to high school in Carleton Place, my family has owned land here for more than fifty years, and the county has awarded me a heritage prize for my work on protecting wild places. No one is paying me to write this article. I am writing simply because this is my home and because the wild creatures who live here cannot speak for themselves.

So, with this background, and cautions, here is my provisional list for some of our priority natural areas in Lanark County.

One final caution. The list is not complete. There are other green gems that could be added. For example, the list does not include all land trust properties. Nor does the list include all of our provincially significant wetlands, only the larger ones. A refined and complete list is needed. I look forward to our county requesting its planner to hire a biologist to prepare a comprehensive document for our county, with a complete list of green gems, proper documentation for each one, and a map of each area. This short article is just a beginning. It is not a full plan.

Green gems. Each green gem could use some friends who care about its future. We need to protect these areas for the benefit of all the wild creatures that live there. We need to protect these natural areas so we don't end up with a depressing landscape of urban sprawl like so much of southern Ontario. We also need to protect these areas for the benefit of our grandchildren and for future generations. We also need them for ourselves, just so we can have the pleasure of seeing ospreys diving to catch a fish or hearing wild loons call at dusk. (This appendix was published as an article in *The Lanark Era* in 2019.)

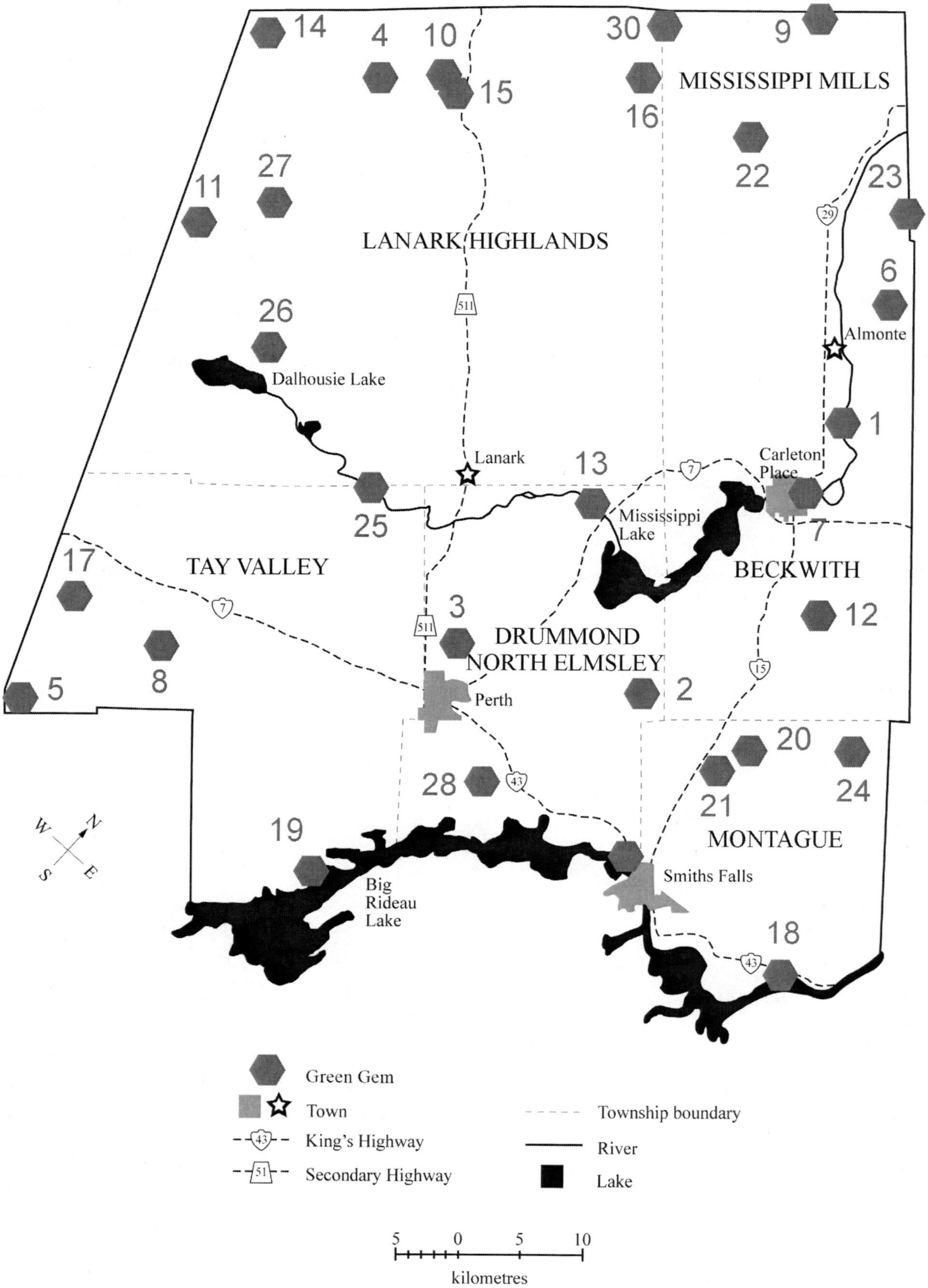

Map 20. Green Gems of Lanark County.

Green Gems of Lanark County

This is a provisional list of the green gems of Lanark County—areas with our most important natural landscapes and highest concentrations of wild species. Each needs to be protected at the county and township level, with a core area of natural habitat surrounded by a buffer zone with limited human development. This list is compiled from information collected by professional biologists. Areas in bold include significant amounts of upland, that is, areas not protected by wetland regulations. The primary sources are given in brackets[1,2].

1. Appleton Silver Maple Swamp (candidate ANSI, NELC, PSW)
2. Black Creek Wetland (PSW)
3. Blueberry Marsh (ANSI, NELC, PSW)
4. Blueberry Mountain (protected by Mississippi Madawaska Land Trust) (PLC)
5. Bolingbroke/Bob's Lake deciduous woodlands (PLC)
6. Burnt Lands Alvar (ANSI, NELC, PLC)
7. Carleton Place Hackberry Forest (NELC)
8. Christie Lake Fire Barrens (candidate ANSI, NELC, PLC)
9. Clay Bank Alvar (PLC)
10. Darling Township Forest (includes Napier Lake) (ANSI, NELC, PLC)
11. Dixon and Bottle Lake Shoreline Wetlands (PLC)
12. Goodwood Marsh (candidate ANSI, PSW)
13. Innisville Wetlands (ANSI, PSW)
14. Lavant Long Lake (NELC, PLC)
15. Lavant/Darling Spillway (NELC)
16. Lowney Lake (part of candidate ANSI, PLC, part of PSW)
17. Maberly Bog (candidate ANSI)
18. Merrickville Marsh (candidate ANSI, PSW)
19. Murphys Point Provincial Park (NELC, PLC, PSW)
20. North Montague Swamp (candidate ANSI, PSW)
21. Numogate Mud Lake Fen (ANSI, PSW)
22. Pakenham Mountain (NELC, PSW)
23. Panmure Alvar (candidate ANSI, NELC, PLC)
24. Pinery Road East Alvar (PLC)
25. Playfairville Rapids (NELC, PLC)
26. Purdon Conservation Area (NELC, PLC)
27. Robertson Lake (PLC)
28. Tay Marsh (candidate ANSI, PSW)
29. The Swale Marsh (candidate ANSI, PSW)
30. White Lake Wetlands (candidate ANSI, NELC, PSW)

[1] ANSI (Area of Natural and Scientific Interest documented by the Ontario Ministry of Natural Resources and Forestry (Map 16, page 49), NELC (Special Place described on pages 50-60), PLC (Significant Area documented by David White in Plants of Lanark County, Ontario. www.lanarkflora.com/areas.html), PSW (Provincially Significant Wetland designated by the Ontario Ministry of Natural Resources and Forestry based on systematic inventory by teams of biologists)

[2] Note that this list does not contain all 47 provincially significant wetlands in Lanark County, only the largest. These wetlands are each individually important. Each has been evaluated by teams of expert biologists and is protected by provincial regulation. You can view more significant wetlands in Map 14, page 41.

Appendix B

Priority Steps for Meeting 21st Century Conservation Challenges

I am often asked the same question: "What can we do to protect Lanark County for the well-being of future generations?" The first step is to know what we have in the county, and to appreciate its significance. That is why I prepared this guide. The second step is to become aware of the threats to our wild places and wild species. Many important natural areas, and significant species have been mentioned by name in this guide. The third step is to take action to improve the situation. Here are specific actions that need immediate attention.

1. Protect all thirty Green Gems as special places.

    Many of these Green Gems (Appendix A) still have minimal legal protection, or none. In addition to proper protection, all need surrounding buffer zones.

2. Protect all natural forests on crown land in Lanark County.

    Currently the forests and other wild landscapes on crown land have been given to the Mazinaw-Lanark Forest Inc (a consortium of "7 sawmills, 12 independent logging companies, and one pulp mill") and are scheduled for continued logging.

3. Complete a county-wide environmental plan.

    The plan needs to show core natural areas, buffer zones, and environmental corridors that will connect them.

4. Complete environmental plans for each township in Lanark County.

    These should describe and map the core natural areas in each township, as well as show buffer zones, and the environmental corridors that will connect them. These township level plans could fit together to comprise the county wide plan.

5. Enhance carbon storage to address climate change.

    Protect all county wetlands and forests from any activities that will reduce their capacity for carbon storage.

6. Species at risk that live in Lanark County need to have their habitat and populations protected.

    Critical habitat for endangered species should be indicated in township and county plans, and protected within core areas, buffer zones, and environmental corridors.

Made in United States
Cleveland, OH
04 November 2026

47646340R10047